Building Chaos

Speed is very much the essence in today's construction industry, both for its business leaders and its workers. However, in this brave new world of rapid response and impermanent specialization, when is flexibility a supple and fluid response to changing circumstances, and when is it a frightening and chaotic bouncing from one crisis to another? In the face of turbulent and unreliable demand, what is to be done to ensure projects reach a safe and profitable conclusion?

This important collection of case studies shows how nine countries have each addressed these questions through the regulation or deregulation of their construction industries. These essays contrast the experience of the more highly regulated construction industries (those of the Netherlands, Germany and Denmark) with primarily deregulated industries (Korea, the United Kingdom and the United States of America) and those countries whose industries fall in between these two extremes (the Quebec province of Canada, Spain and Australia).

This clearly written, well-argued book is an extremely useful volume not just for students of industrial organization, work organization and labor economics, but also to all those involved with the construction industry, whether it be contractors, construction workers, labor representatives, consultants or policy-makers.

Gerhard Bosch is Vice President of the Institute for Work and Technology, Professor of Sociology at the University of Duisberg.

Peter Philips is Professor of Economics at the University of Utah.

Industrial Organization/Labor Economics

Routledge studies in business organizations and networks

Building Chaos

An international comparison of deregulation in the construction industry

Edited by

Gerhard Bosch and Peter Philips

Routledge
Taylor & Francis Group

LONDON AND NEW YORK

First published 2003
by Routledge
11 New Fetter Lane, London EC4P 4EE

Simultaneously published in the USA and Canada
by Routledge
29 West 35th Street, New York, NY 10001

Routledge is an imprint of the Taylor & Francis Group

© 2003 editorial matter and selection, Gerhard Bosch and
Peter Philips; individual chapters, the contributors

Typeset in Goudy by
Integra Software Services Pvt. Ltd, Pondicherry, India
Printed and bound in Great Britain by
Antony Rowe Ltd, Chippenham, Wiltshire

British Library Cataloguing in Publication Data
A catalogue record for this book is available from the British Library

Library of Congress Cataloging in Publication Data
A catalog record for this book has been requested

ISBN 0–415–26090–6

Contents

Figures

Tables

Contributors

Gerhard Bosch is Professor of Sociology at the University of Duisburg and Vice President of the Institute for Work and Technology, Science Center North Rhine–Westphalia in Gelsenkirchen. He is an expert on labor markets and has published several books in German, English, and French. He has focused on the future of work, redundancy, training, working time and the service economy. He has been an advisor for the EU, the ILO, and the OECD. He is the author of *Retraining not redundancy: Innovative approaches to industrial restructuring in Germany and France* and, with Klaus Zühlke-Robinet, *Der Bauarbeitsmarkt: Soziologie und Ökonomie einer Branche*.

Justin Byrne is a Researcher at the Center for Advanced Study in the Social Sciences, Instituto Juan March, Madrid, and Associate Professor at New York University in Madrid and the Vassar-Wesleyan-Colgate Program in Spain. He was awarded a Ph.D. from the European University Institute, Florence, for his thesis on *Work, organization and conflict: the bricklayers of Madrid, c. 1870–1914* (1998) and is the author of a number of articles on the Spanish building industry past and present.

Jean Charest (Ph.D. Industrial Relations, Université Laval) has been Associate Professor at the School of Industrial Relations of the Université de Montréal since 1997. His research and publications concern public policies on the labor force, industrial relations institutions, labor force training, and unionism.

Mark Harvey is Senior Research Fellow at the ESRC Centre for Research in Innovation and Competition at the University of Manchester. He is contributing to the development of comparative and historical economic sociology in the study of varieties of capitalism. Using an "instituted economic process" (Polanyi) approach, he is researching production, distribution, exchange, and consumption in construction, biotechnology and bioinformatics, and food systems.

Byung-Goo Kang is Assistant Professor at the Department of Economics of Inha University in Korea. He earned a Ph.D. in Economics from the State University of New York at Binghamton. His interests are in the area of social safety net programs, and his publications include *Public Works Project as* *a Social Safety Net in Korea, Beyond the Restructuring Policies in Korea*, and *An Evaluation of Unemployment Measures in Korea*.

Nikolaj Lubanski is Associate Professor at the Danish School of Public Administration and Assistant Research Professor at FAOS, Employment Relations Research Centre. He earned a Ph.D. in International Employment Relations at the University of Copenhagen. His research focuses primarily on comparative studies of labor market development in Europe, theories of labor market relations, and the sociology of work. He is the author of *The Europeanisation of the Labour Market* (1999) and *The Impact of Europeanisation on the Construction Industry: A comparative analysis of developments in Germany, Sweden and Denmark* (1999).

Peter Philips is Professor of Economics at the University of Utah and the co-editor of *Three Worlds of Labor Economics* (1986) and co-author of *Portable Pensions for Casual Labor Markets: The Central Pension Fund of the Operating Engineers* (1995). He has published widely in academic and industry journals on the canning and construction industries. He has been a consultant for the US Labor Department and has worked as an expert on the Davis-Bacon Act for the US Justice Department.

Elsa Underhill is Senior Lecturer in Industrial Relations at Victoria University, Melbourne. She has published upon contract labor, self-employment, and portable employment benefits in the Australian building industry, in a number of journals and monographs, and was a co-editor of a special edition of the journal *Labour & Industry* on precarious employment in Australia. Her other publications deal with human resource and industrial relations strategies in the Australian steel industry (with D. Kelly), deregulation of the Victorian labor market, labor hire employment, and occupational health and safety management systems (with C. Gallagher and M. Rimmer). In 1995 she was President of the Association of Industrial Relations Academics of Australia and New Zealand.

Marc van der Meer is Assistant Professor at AIAS, the Amsterdam Institute for Advanced Labour Studies, at the University of Amsterdam. His doctoral dissertation discussed collective bargaining and the allocation of employed, new entrants, and unemployed in the construction industries of Amsterdam and Madrid (1998). Currently he is involved in research on innovation in collective bargaining in the Netherlands and in European labor market research on segregated labor markets.

Jin Ho Yoon is Professor of Economics in the Division of Economics at Inha University and a Visiting Scholar at the Department of Urban Studies and Planning at MIT. He graduated with a Ph.D. in Economics from Seoul National University in Korea. For 20 years his research has focused on labor economics and industrial relations. He has contributed to several government committees, including the Tripartite Commission, which consists of the representatives from labor unions, employers, and the government, and has been a frequent advisor to labor unions. His publications include *Working Time and Work Organization in Korea, IMF Bailout and Employment Crisis: The Labour Response, Neo-corporatism in Korea?: An Evaluation of the Tripartite Commission, The Economic Crisis and Working Time Issues in Korea*. His recent paper, "Working Time and Work Organization in Korea," will be published by the International Labour Organization.

Klaus Zühlke-Robinet is a scientific employee at the project management organization (work layout and services) of the Federal Ministry for Education and Research in Bonn. Previous to this he worked several years in the Institute for Work and Technology in Gelsenkirchen and was, amongst others, working on the project *Der Bauarbeitsmarkt: Soziologie und Ökonomie einer Branche* (together with Gerhard Bosch). Furthermore, his articles regarding bad-weather pay, professional training, branch mobility, and the Europeanization of the construction industry in Germany have been published in various journals.

Acknowledgments

This volume was made possible in part by the Center to Protect Workers' Rights (CPWR) as part of a research agreement with the US National Institute for Occupational Safety and Health (NIOSH) grant #U60/CCU317202-02-1. The research is solely the responsibility of the authors and does not necessarily represent the official views of NIOSH or CPWR. CPWR is a research and development institution associated with the US Building and Construction Trades Department, AFL-CIO. CPWR, Suite 1000, 8484 Georgia Ave., Silver Spring, MD 20910; Tel. 301-578-8500; Web http://www.cpwr.com.

The Hans Böckler Foundation in Düsseldorf and the Ministry for Urban Planning, Housing, Culture and Sports of North Rhine–Westphalia in Düsseldorf were responsible for the financial support of the conference and project meeting out of which this book developed. That conference, "Labor markets in the construction industry. An international comparison," was held in Gelsenkirchen on 19–20 October 2000. Hans-Böckler-Stiftung, Hans-Böckler–Str. 39, 40476 Düsseldorf, Germany; Tel. (0211) 77 78-0; Web http://www.boeckler.de. Ministerium für Städtebau und Wohnen, Kultur und Sport des Landes Nordrhein–Westfalen, Elisabethstraße 5–11, 40217 Düsseldorf, Germany; Tel. (0211) 38 43-0; Web http://www.mswks.nrw.de.

Many thanks also go to Sandra McIntyre for her outstanding editing work on this volume. Editing was equally financed by the Center to Protect Workers' Rights and the Institut Arbeit und Technik, Wissenschaftszentrum Nordrhein–Westfalen, Munscheidstraße 14, 45886 Gelsenkirchen, Germany; Tel. (0209) 17 07-0; Web http://www.iatge.de.

1 Introduction

Gerhard Bosch and Peter Philips

THE FOCUS OF OUR STUDIES

Jack be nimble! Jack be quick! Such advice is the order of the day for construction companies exposed to a world of intensified competition. This also is sound advice for many construction workers, who may face the prospect of moving from employer to employer and even from occupation to occupation over their work lives. However, in this brave new world of rapid response and impermanent specialization on the part of both firms and workers, when is flexibility a supple and fluid response to changing circumstances; and when is it a frightened and chaotic bouncing from one crisis to the next? How can firms accumulate intellectual and physical capital, suitable to the problem at hand, if the firm must be ready to re-gear for an entirely new and unknown problem in the very near future? Once acquired, how can the accumulated human capital be retained in the face of fluctuating work? How can workers acquire the human capital specific to the problem at hand if the problem keeps switching? Finally, even with a valuable stock of skills, what allows the worker to remain in the construction industry if it offers only uncertainty and insecurity? In short, in the face of turbulent and unreliable demand, how can the associated risks be mitigated so that both human and physical capital can be accumulated, experience can be gained, and the problem at hand can be done efficiently and productively?

The construction industry has had to be nimble and quick throughout its modern history. The construction process is an inherently turbulent process, with the turbulence of production rooted in the fact that construction products are durable goods susceptible to postponable demand. Demand is postponed each time the business cycle turns down. Although construction products are durable, they are neither easily stored nor easily transported. A non-storable, non-transportable durable good is especially sensitive to the business cycle. Unwanted buildings cannot be shipped elsewhere where demand is still vibrant. Furthermore, buildings and roads are difficult to stockpile in anticipation of future demand. Thus, the nature of the product prevents it from smoothing out the peaks and valleys of demand. As outdoor work, construction is also subject to seasonal swings, which add to the turbulence of the industry. Since construction provides geographically specific products, uneven development by region again adds to the industry's turbulence. Consequently, the problem of efficiently deploying and re-deploying labor and resources in construction is both chronic and acute. The geographic, seasonal, and cyclical instability of construction adds significantly to the risks associated with accumulating physical and human capital within the industry.

Not every country manufactures automobiles nor mines coal. However, every country has a construction industry. Indeed, virtually every region and locality in every country has the capacity to build houses, commercial buildings, and roads. This book is about an industry that is found everywhere. In broad terms, construction employment will account for 5–10 per cent of almost every local labor market in every country in the world (see Table 1.1). This book is about nine of those countries – five in Europe (the Netherlands, Germany, Denmark, Spain, and the United Kingdom), two in North America (the United States and the province of Quebec, Canada), and two in the Far East (Australia and South Korea). The authors from these countries will describe the general characteristics of each of these particular construction industries and will focus on their construction labor markets.

In our research, we have found that construction industries in these countries develop along two very different paths. These paths diverge based on how, if at all, the long-term costs of construction are paid for. These long-term costs include the training of the next generation of construction workers, the maintenance of the health and safety of the current construction workforce, the compensation of workers for the instability of the industry, and the support of the past generation of construction workers during the years after their retirement from the industry. These long-term costs also include the down-stream maintenance costs of construction projects and the development and incorporation of new technologies into both the construction process and the construction product.

Table 1.1 Share of the construction industry of total employment, in per cent, 1988 and 1998

Country	1988	1998
The Netherlands	6.12[a]	6.23
Germany	7.31[b]	7.88
Denmark	6.25	6.02
Canada	6.82	6.07[c]
Australia	7.51	7.29
Spain	9.12[d]	9.31
United States	4.85	4.78
United Kingdom	–	–
Korea	7.44[e]	7.91

Source: OECD; United States Bureau of Labor Statistics.

Notes
a = 1995; b = 1991; c = 1995; d = 1995; e = 1990; – = no data available.

Because construction is a highly seasonal and cyclically volatile industry, often there is high labor turnover within the firm and in the industry. Often there is also relatively rapid turnover of the firms themselves within the industry. Again, this turbulence makes the accumulation of human and physical capital problematic. The organization of construction through a system of articulated subcontracting can make skills formation and capital accumulation even more difficult. As we shall see, the competitive structure of articulated subcontracting in construction makes it difficult to include in short-term bids the long-term needs of the industry. Given the turbulence, organizational structure, and competitive pressures of the industry, institutions, customs, and regulations become crucial determinants of the type of development path a given construction industry follows. Indeed, a major purpose of this book is to compare and contrast the distinct institutions, customs, and regulations (and de-regulations) of nine countries to see how these differences respond to, and shape the underlying potential for turbulence and uncertainty in construction.

One growth path for construction industry development is capital-intensive, human-capital-intensive, and technically dynamic. Along this path, the long-term costs of the industry – the cost of training, the cost of maintaining the health and safety of the workforce, the costs for compensating workers for job instability, the cost of maintaining retired construction workers in their old age – are sometimes financed by levy systems arranged through the State or by the industry through collective bargaining between unions and employers or other industry-focused regulations. At other times, these long-term costs are covered through society-wide government programs or regulations. In countries with such a "high road" strategy, the industry is able to build up a stable core workforce.

The second type of construction industry development is along a low-skill, less well-equipped, labor-intensive growth path. Construction industries on this path have more difficulty incorporating new technologies and developing modern forms of work organization. These low-wage construction industries have trouble creating and retaining a skilled, craft labor force. Many of the long-term costs and needs of these construction labor markets go unmet or are shifted to groups with the least ability to bear these costs. In countries with such a "low road" strategy, the stable, core workforce is very small, while instability among the majority of workers is quite high.

The path chosen will depend on the institutions, regulations, and customs that shape how the problem of "free riding" is handled. We shall find that the construction industry is rife with free rider problems. Under the short-term competitive pressure to win bids, construction contractors are reluctant to place into their bids, costs that may help sustain the long-term health of the industry but that do little to achieve the immediate goal of undercutting a competitor's bid. Rather than paying for the cost of training, a contractor's best competitive strategy may be to try to take a free ride by using the labor force trained by someone else. Unfortunately, when every contractor thinks this way, they fail to train anyone adequately. Where contractors fail to train, workers have the option of training themselves. However, in a volatile industry where continued employment prospects are uncertain, workers may choose not to invest in construction-specific skills for fear of idling or losing that investment. In this risky industry, as pressures shift the focus of human capital investment from employers and the industry to individual prospective workers, skill investment activity is left to those least able to make that investment. This shift in investment responsibility from the firm or union to the worker also places investment decisions in the hands of individuals with less ability to diversify their investment simply because it is embodied in themselves. This increases an already risky endeavor. Shifting human capital investment from the firm, the labor organization, the industry, and/or the State to the individual prospective worker reduces the amount of skill formation that will take place in construction.

However, where human capital formation is inadequate, the process snowballs. Contractors become reluctant to entrust expensive equipment to a less trained workforce. Less trained and less well-equipped workers have difficulty incorporating new, specialized, and advanced technologies into the structures they build. Furthermore, a less trained construction workforce does not develop lasting ties to the construction industry. Having little in the way of industry-specific human capital to lose, a less trained workforce more easily slips away when the downturn comes. Thus, not only do skills not accumulate, but also industry-specific work experience is cut short. In the long run, limited construction capabilities due to lack of training and experience limit the industry's ability to build quality, technically advanced structures safely. This can come back to haunt the construction industry itself. Construction industries everywhere are dependent upon the health of the local economy. In a world economy where the capacity to generate and adapt to rapid technological change is a key to local success, local construction industries that are not capable of building and maintaining a reliable, technically up-to-date infrastructure can handicap overall local economic growth.

The purpose of this book is to describe these two paths along which the construction industry has developed. Based on these nine case studies, we ask how these labor markets function and what public policies, regulations, market contracts, and customs foster development along the high-skill, high-quality growth path, and what policies and approaches foster the low-skill, cheap construction growth path.

A GENERAL FRAMEWORK FOR UNDERSTANDING THE CONSTRUCTION INDUSTRY

Characteristics of the construction industry

While the thrust of these case studies is to show that construction labor markets can be organized in very different ways depending upon the legal, regulatory, organizational, and cultural environments of each country, there are, nonetheless, certain core characteristics shared by all construction industries. These core characteristics are rooted in the product itself.

Construction industries build durable products for government, industry, commerce, and households. Once built, a road is meant to last for many years indeed. Factories, stores, office buildings, and homes have expected lifetimes that can range from 10 to 100 years. Construction products are among the most durable of the goods produced by capitalist economies. Ironically, the durability of its products makes construction one of the most volatile and cyclically erratic of industries. In a few instances, the demand for this durable product comes from a government policy to buy public goods regularly. However, the much more usual case in capitalist economies is for the demand for construction goods to come from the more erratic demands of producers and households, often responding sharply to changes in interest rates or future expectations.

Commercial and industrial demand for durable construction products is rooted in future expectations of profitable enterprise activity. This varies exponentially with the business cycle. Excited business people fall over each other trying to bring on line new commercial and industrial capacity during the boom. As the economy slows down and expectations diminish, prospective construction is often among the first of downstream projects to be put on hold. Similarly, residential demand for new homes and apartments is highly sensitive to interest rates, tying this type of construction as well to the business cycle. Government demand for construction can likewise be tied to, and exacerbate the business cycle, either by downsizing expenditures as tax revenues dry up in the downturn or by mishandling efforts at counter-cyclical monetary and fiscal policies. In most cases, when the economy gets a cold, construction gets the flu.

Thus, construction is like the subsectors of manufacturing that produce durable products. However, construction is also like agriculture. Both are primarily or substantially outdoor activities subject to the vagaries of weather. This is entirely true of road construction. In building construction, eventually the new structure takes shape and the work comes indoors. Nonetheless, construction work is affected by the weather and employment varies with the season. In more temperate climates, there is less seasonal fluctuation in construction activity, while in harsher climates construction work can be quite seasonal.

Construction also shares a crucial characteristic with most activities in the service sector. Like most service work, productive activity in construction takes place at the point of purchase. In manufacturing and agriculture, the good is made or the crop is harvested, and then the product is shipped to a sales location and sold to the customer. In many service activities – getting a haircut, watching a play, riding a bus, receiving an education – the production of the service takes place in the presence of the customer. Construction also must travel to the customer before commencing work. It is not possible to build a road somewhere else and ship it to the place where it is needed. While modular homes are built on a manufacturing basis and then shipped to the customer, most buildings must be built on site. This is why every place has a construction industry. Every locale must have at least some of the capabilities to produce locally the construction products needed in that area. While construction is usually referred to as a goods-producing industry, the fact that construction typically takes place where the customer needs the product explains why the industry is often referred to as "construction services." For the purposes of studying the construction labor market, the service aspect of construction is important because it adds to the instability of employment. In this case, the instability is neither cyclical nor seasonal but rather the instability of moving work from place to place in search of new customers.

Sometimes new durable construction goods can be prefabricated and installed on site. As will be noted in the US case, as much as one-third of new residential housing is composed of prefabricated units, built in a manufacturing process and then shipped in segments to be assembled and placed on a foundation at the customer's location. The process of prefabrication need not be this dramatic. For example, while cabinets and cupboards were historically built on site in new homes, today it is quite common for cabinets and cupboards to be prefabricated in a manufacturing setting, then shipped and installed at a new home site. Electrical fixtures and doors hung in their frames are other examples of prefabrication where what was once construction work on site is now manufacturing activity shipped prefabricated to the construction site. From the perspective of the construction labor market, prefabrication is a process of transferring activity from construction to manufacturing. Prefabrication may or may not be a labor-saving technology, but it surely is a labor-displacing technology from the perspective of the construction work site. Prefabrication in construction is somewhat similar to a runaway shop in manufacturing. They both move work away from its present location.

Historically, there has been a continuous process of moving work away from construction into manufacturing through innovations in prefabrication. This process of change focuses on making construction materials semi-finished products. A second form of technological change focuses on equipment. The construction industry in most countries has actively adopted labor-saving equipment. For example, the size and capacity of road-building and earth-moving equipment have grown dramatically. Carpenters may now use nail guns rather than hammers. Yet, historically, while manufacturing employment has fallen as a share of the overall labor force in many advanced capitalist economies, when one looks past cyclical swings in construction employment, construction's share of overall employment in all countries and most localities remains relatively stable. Why is this?

Two reasons account for the relative permanence in the local importance of the construction labor market. First, through prefabrication, a company can take parts of the construction process and move them to a manufacturing plant elsewhere. However, while the mobility of manufacturing plant and equipment is a strong force for reducing or eliminating manufacturing labor from a locality, prefabrication is only a weak force in removing construction labor from a community or economy. Second, labor-saving devices are typically easier to apply to most manufacturing processes compared to construction. Construction products tend to be non-uniform; construction sites differ in terrain and buildings differ in their design. While vigorous efforts are made to standardize building products and their components, construction projects resist standardization to a greater degree than manufactured products. Thus far, customers have also resisted construction standardization to a greater degree compared to standardization in manufacturing. For instance, however wide the range in automobile brands and styles, the range in residential housing is wider. The production of customized construction products limits the ability of labor-saving mechanization to shed labor from this industry. So, on the one hand, construction employment is uniquely erratic due to the seasonal and cyclical fluctuations of this industry. On the other hand, construction employment shows more long-term relative stability compared to manufacturing because there are limits to the extent to which construction work can run away through prefabrication or be displaced through mechanization. Furthermore, the customized character of construction may limit the degree to which the craft nature of construction work can be reduced through de-skilling.

The customized character of construction work may add to the dangers of this work. Each site is new and in some ways unfamiliar. The routine of work cannot be as easily standardized and thus made more certain as in the case of manufacturing. Indeed, construction contains some of the characteristics of transportation and mining that make those industries unusually dangerous. Construction activity necessarily entails moving heavy materials into place. The activity of trucks, cranes, earth moving equipment, and other heavy equipment on road and building construction sites brings to construction many of the dangers of transportation. The work of trenching, the digging of foundations and basements, and the movement of earth in road and dam construction bring to the work site many of the dangers of mining. In addition, construction has its own unique dangers. Construction work is often done in high places. The work is often loud, sometimes leading to hearing loss, and often exposed to the elements. Every new work site presents some unique problems and unforeseen dangers. By its nature, construction activity exposes workers to the dangers of slipping, falling, or being crushed by falling materials. Workers are exposed to electrical hazards and toxic substances.

Thus, construction shares something with each of the other five main areas of every economy. Construction employment is seasonal like agriculture; it is cyclical like other durable goods in manufacturing; it is local like many services; and it is unusually dangerous like transportation and mining.

The role of subcontracting in construction

Construction is an unusually risky endeavor, and work site dangers form only part of the risk. Employers who risk capital in construction face considerable financial risks. As we have seen, construction is highly volatile. Responding exponentially to the business cycle, the demand for construction services booms and collapses unpredictably. Demand flits about from place to place as regional business cycles differ, and as jobs start up and shut down in any one locale. Demand can even fluctuate within a locale between various sectors of construction. How construction organizes itself to face, avert, minimize, or shift these risks varies considerably across countries. However, one common response to the risks faced by capital in construction is the use of subcontracting systems to share and shift risk.

Typically, construction activity is divided into four broad sectors – residential housing, commercial buildings, industrial facilities, and infrastructure construction (roads, dams, pipelines, power and communication transmission lines, etc.). This last sector is called "civil engineering" in the European context and "heavy-and-highway construction" in North American terminology. Each of these sectors of construction activity follows its own somewhat distinct business and seasonal cycles. Regional and sectoral differences in an already volatile seasonal and cyclical business present a moving and difficult target of opportunity for profit. Capitalists run the real risk of idling their fixed capital investments by missing the next target of opportunity in the ever-shifting structure of construction demand. All capitalist construction industries in all countries have devised strategies to limit exposure to the risks of idle capital when the demand for particular construction services in local areas collapses. While strategies vary considerably from country to country, they all share one common characteristic. As most construction firms are small with relatively little capital and relatively few employees compared to other goods-producing industries, there is limited capital put at risk in the face of unusual turbulence and uncertainty.

How can large construction jobs be done by an industry characterized by small firms? Everywhere in developed and newly developing capitalist economies, the answer is found in articulated subcontracting. Construction firms divide into types based on where they place themselves in a system of subcontracting. General building contractors bid for and take on commercial and residential building projects. They, in turn, subcontract out part or even almost all of these building projects to specialty subcontractors who agree to do special aspects of the construction project. A second set of general contractors bids for, and takes on industrial, road, dam, and other heavy construction. These civil engineering contractors (or heavy-and-highway contractors) then subcontract some or almost all of the work they have won to subcontractors who may subcontract the work even further, to additional layers of subcontractors. The extent of subcontracting varies considerably from country to country, as we shall see. In addition, small construction companies or individuals working outside the subcontracting systems do small jobs of maintenance, repair, additions to existing buildings, and very small new construction. Moreover, it is often common for distinct architectural firms and civil engineering firms to provide design and engineering services needed to plan the construction project. Project management firms may also contract with the final owner of the building project to assist the owner in purchasing the various contracts needed to set a construction project in motion. These project management firms may also play a role in overseeing the construction project or evaluating the project at completion.

[handwritten marginalia at top: not all regulation is bad + all regulation is not a good, in + of itself. there needs to be a higher purpose served by any do reg?]

Also, construction contractors themselves may seek to provide some of these architectural, engineering, and management services. Construction equipment companies may also get into the act by renting equipment to construction contractors.

The prevalence of architectural, engineering, and project management firms in this industry underscores the very different role played by the consumer of construction services, when compared to the consumers in many other markets. While consumers of residential housing sometimes choose from an array of new homes designed by the builder, in custom home construction and most non-residential construction, consumers usually play an active role in designing the product they purchase. The complexity of construction products requires that these consumers hire architectural and engineering services to help them with conceptualizing and implementing their design. Not only do most construction consumers need help designing what they want, but also they often seek help in purchasing what they design. Because construction goods are durable, many consumers of construction services only seldom enter the market. When they do, they are confronted with a potentially bewildering system of subcontracting and material providers. Project managers present themselves as professional buyers of construction services capable of assisting the consumer in the purchasing of not only construction services but also architectural and engineering services.

The activist role of the construction consumer adds to the complexity of the industrial structure of the industry. Contrast this structure to that of automobile production. The automaker designs, makes, and sells cars. The construction contractor does not design the building or road in most cases. The idea of the building has been sold to the customer before the construction contractor bids on the project. Moreover, the winning general contractor does not even make most of the contracted product. That is subcontracted out to others. Indeed, the general contractor in some cases will be pushing actual construction work down towards subcontractors while at the same time attempting to obtain some of the work associated with the design or engineering of the project. All the while, the project manager may be attempting to assume some of the site management work traditionally associated with the general contractor. Thus, the construction industry presents a constantly evolving structure of who does what. The relative roles of consumer, designer, manager, and "constructor" are constantly shifting. This constant evolution of structure is both a source of and an attempted – if temporary – solution to the ever-present turbulence of this industry.

Customized production creates risks. There is the risk customers will not buy what they actually want or need. There is the risk that, what customers buy will deteriorate unexpectedly. The burden and responsibility for these risks is shared and shifted among the various entities that conceive, design, purchase, and construct the customized products of the construction industry. Thus, the evolving structure of construction responds not only to the turbulence of the industry but also to the uncertainties associated with buying and making construction products.

The structure of subcontracting varies across countries and across construction subsectors within countries. Governmental policies play an important role in the structure of subcontracting. This can happen directly through contractor licensing regulations and indirectly through a host of policies and building codes that exacerbate or moderate the risks construction firms and customers face. Tax policies can also play a role. In some cases, the structure of subcontracting can be articulated to the point that a contractor treats many individual workers on a construction site as self-employed subcontractors to avoid paying the taxes associated with wage work.

The structure of subcontracting on the construction site can be divided into two types – cooperative, and competitive subcontracting. In cooperative subcontracting, each subcontractor is assigned distinct tasks over which that subcontractor has sole responsibility. In competitive subcontracting, two or more subcontractors are assigned the same task and their longevity on the job is dependent upon their ability to out-compete their rivals.

Subcontracting is a method for managing risk, but it also can be a method of labor control and a way to cut labor costs on construction work sites. In manufacturing, there have been two widely used methods for managing labor costs. The first is extending the detailed division of labor into ever more finely defined tasks that require fewer skills to perform and are more easily monitored because of their simplicity. The second method is tying workers to an assembly line where a worker's job is defined by a carefully laid-out progression of tasks that are typically mechanically interconnected. In this case, the speed of work is roughly determined by the speed at which the assembly line is run. While construction work is not amenable to assembly line production, highly articulated subcontracting systems can serve to introduce and extend specialization within the division of labor on the construction site. This is the case in cooperative subcontracting. However, this is not always the purpose or result of subcontracting. As mentioned above, risk shifting and risk reduction provide a second reason for extending the subcontracting system. Competitive subcontracting does not increase specialization; rather it introduces competition on the job site among similar subcontractors. The articulation of subcontracting down to individual workers is usually designed to evade regulations protecting wage workers rather than designed to extend specialization. These competing reasons for articulating subcontracting may work against each other. For instance, the cost-cutting goals of competitive subcontracting and independent individual subcontractors may heighten the risks general contractors face in attempting to meet construction deadlines and quality standards, because they come to rely on low-skilled workers.

The role of regulations in construction

Regulation of the construction industry varies widely from country to country. It is beyond the scope of this volume to explain why one country adopted one set of regulations while another country has a very different set of rules; rather, we will focus on the effects of differing regulations and standard practices.

In general, government rules and industry practices – often the result of collective agreements between unions and employers' organizations – help shape the risks that construction firms face and the nature of the competition these firms experience. An example of how regulations can change the face of the construction industry is found in country variations in the way construction workers are trained.

The skills and knowledge that a construction worker acquires, through either formal training or on-the-job experience, may be thought of as a form of investment in human capital. The time and expense involved in creating these skills is the investment, and the higher productivity and quality of the worker's labor effort comprise the payoff. Workers have as much trouble accumulating human capital in construction as contractors have in risking fixed financial and physical capital. The volatility of construction demand in countries – where volatility is not dampened by government spending, nor mitigated by bad-weather allowances, nor offset by generous unemployment benefits – leads to extended spells of punishing unemployment. Human capital can be idled just as easily as physical capital. Unmitigated volatility and unemployment present the prospect of idled human capital and stand in the way of skill formation.

Construction contractors who want skilled workers must pay a double premium to induce workers to invest in themselves. They must pay a reasonable return for that investment and an additional premium to offset the risks of unemployment. However, even premium construction wages might not bring forth workers, ready to invest in themselves. Those who would become construction workers tend to come from working class families that have not acquired considerable discretionary funds for human capital investment. Furthermore, loans for training construction workers are not widely available. In an industry characterized by high unemployment, the risks of default are great, and bankers have difficulty repossessing loans for education. Moreover, contractors themselves may be reticent to invest in the training of workers. Volatile construction demand means turbulent construction labor markets. Most workers in unregulated construction labor markets are unlikely to stay with typical contractors for extended periods. Contractors are loath to spend money on the training of a worker who may soon be the employee of a competitor. Finally, customers are reluctant to pay for training as well. Since construction projects create durable goods, many customers come to market only rarely. Training in construction takes time – usually longer than the duration of any construction project. Thus, what these customers need are trained workers today, not a trained workforce tomorrow. These customers rely, therefore, upon past training and, except for some project-specific task training, are reluctant to pay the incremental cost of training the future workforce. So, in general, considerable short-term pressures and considerations militate against the accumulation of human capital in construction.

In the absence of regulations, therefore, construction tends towards a path of small firms with limited fixed capital and unskilled workers with limited human capital. Subcontracting knits this unregulated system together to allow these small firms to take on big jobs. However, it remains problematic whether the human capital will be found in the labor pool to finish the job. Extended subcontracting can reduce the need for human capital by introducing to some extent a detailed division of labor that reduces skill requirements. On the other hand, competitive subcontracting and independent individual subcontracting can create worsening job conditions, speed-ups, and stretch-outs that undercut the possibilities for training through worker fatigue. Ultimately, the customized nature of most construction, the dangers of construction work, and the risks associated with assuring the quality of the construction product combine to impose irreducible needs for some skills within the construction labor force – i.e., skills that cannot be eliminated through the market solution of articulated subcontracting.

Regulations overcome these shortcomings in a wide variety of ways. Governments sometimes directly finance the training of construction workers. Some governments create licensing requirements designed to force training. Governments sometimes require that training take place on all government construction or tax contractors to force long-term human capital investment. Other governments support or promote collective bargaining or other contracts capable of generating training. In some countries, the social partners are so strong and well organized that they themselves agree upon such regulations and jointly enforce them.

In addition to these labor market regulations, governments can use product market regulations to induce human capital investment. They can impose binding quality standards or require the training of would-be contractors before they may enter or continue in the industry. In imposing these regulations, governments represent the long-term interests of the industry and those of the customers served. Government also has a direct interest in the long-term quality and productivity of the construction labor force. Public entities come to the construction market more often and more continuously than do most private customers. Unlike the private, occasional customer who must take as a given the existing quality of the construction industry, government has a direct interest in seeing that this quality is maintained and developed, for the sake of future government projects as well as the general needs of the economy.

A COMPARISON OF CONSTRUCTION
IN NINE COUNTRIES

The structure of regulation and deregulation is complex and multidimensional. We have for simplicity's sake ordered the case studies within this book along a roughly arranged continuum from more regulated to less regulated construction labor markets. Table 1.2 presents in a most reductionist way our reasons for this ordering.

Four dimensions are considered in ordering construction labor markets from the more to the less regulated. First, is there generally extensive or limited

Table 1.2 Order of case studies from more to less regulated construction labor markets

Country under study	Labor–management cooperation	Collective bargaining	State codification of labor standards	Construction a typical or exceptional case
The Netherlands	Yes	Yes	Yes	Typical
Germany	Yes	Yes	Yes	Typical
Denmark	Yes	Yes	No	Typical
Quebec province of Canada	Yes	Yes	Some	Exceptional
Australia	Yes	Yes	Withdrawn	Exceptional
Spain	Limited	Limited	Limited	Typical
US	Varies by state and trade	Varies by state and trade	On public works in some states	Varies by state and trade
UK	Limited	Limited	No	Typical
Korea	Very limited	Very limited	No	Typical

labor–management cooperation in setting labor and industry standards? Second, is there established and active collective bargaining? Third, does the State engage in codifying and generalizing the results of labor management cooperation and collective bargaining? Finally, are the arrangements in the construction industry and labor market typical of the overall regulatory framework of the economy, or is construction an exceptional case? With these factors in mind, our cases divide roughly into three groups: (a) three basically regulated construction labor markets; (b) three intermediate cases; and (c) three basically unregulated economies.

These rankings are only approximate, and the distinct categories merge to some extent upon closer inspection. For instance, in some trades and states in the United States, the construction labor market is more regulated than the overall construction labor market in Spain. However, in the aggregate, the construction industry in the United States is less regulated than the Spanish construction industry. The devil is in the details. Therefore, we invite the reader to consider the following case studies for a richer, deeper understanding of the nature and effects of regulation on the development, efficiency, and social implications of these nine construction industries.

The Netherlands

Construction is as volatile in the Netherlands as in most countries, yet job stability among Dutch construction workers is relatively high. This is because enterprises hire employees within a labor market structure that allows for permanent, flexible contracts and that provides social funds for vacation pay, bad-weather allowances, vocational training, unemployment benefits, general retirement, early and disability retirement, and other benefits associated with the health and safety of workers.

Marc van der Meer shows that, through a system of state levies and cooperative agreements between construction unions and industry associations, the Dutch construction industry operates within a highly regulated environment. Since 1926, Dutch construction has utilized funds that have been negotiated and set up jointly by labor organizations and employers' organizations. These funds are financed by mandatory levies on contractors and workers to pay for the costs of turbulence within the construction labor market. With social taxes, these funds account for about 45 per cent of an average construction worker's gross labor income. Collective agreements are extended and enforced by the government. The labor market stipulations in these agreements are complemented by additional standards regarding technological requirements and construction quality standards. In general, mandated standards are effectively enforced. Some of this enforcement entails self-policing by industry associations.

This system of regulation has adapted to important recent changes, including increased subcontracting, the development of temporary labor agencies, and the advent of posted workers, by insuring that existing regulations apply to these new aspects of the industry.[1] The Chain Responsibility Act of 1982 made the general contractor responsible for all subcontractors, insuring that these subcontractors pay the required taxes and contributions to social funds. This meant that, in comparison with other countries – notably the UK but also in certain segments of the US industry – extended subcontracting in the Netherlands did not result in black markets nor in a culture of widespread tax evasion. Employers of workers from temporary labor agencies and posted workers are required to pay these workers the same wage rates as Dutch construction workers. The result of these regulations is severe restrictions on free riding with regards to the long-term costs of the industry and on tax evasion strategies with regards to extending the subcontracting system.

Consequently, the Dutch construction labor force continues to be well trained and highly productive. The challenge the Dutch industry faces is to continue to revise and adapt the regulations in this industry to encourage needed change while keeping potential free-rider and tax evasion strategies under control.

Germany

The Handicrafts Code is a regulation unique to the German case that limits access to self-employment in construction to those who become certified masters. As Gerhard Bosch and Klaus Zühlke-Robinet show, this entry barrier has prevented the explosion of self-employment that is found in some other countries, such as in the United Kingdom. At the same time, this regulation has required training among those who are self-employed. As in the Netherlands case, in Germany there are generally binding technical and quality standards, although there are not the quality guarantees and inspections of buildings 5 years after completion as in the Danish case.

In Germany, collective bargaining is more centralized than in any other country in this study. One national agreement covers all the distinct construction trades with respect to contributions into and benefits from various social funds. This agreement is made binding by government enforcement. This unusual centralization of collective bargaining is the result of post-World War II reforms that created a unified construction union and the elimination of traditional divisions by craft and political affiliation. In contrast to unions in other countries, the German construction union does not defend craft demarcations. Training and apprenticeship programs in Germany tend to reduce differences between trades compared to most other training systems. The various social funds for training, bad-weather allowances, etc., are jointly administered by the union and industry groups with about 20 per cent of overall wages going into these funds. Internal labor markets or systems of promotion within the firm are encouraged, and worker ties to the industry are also encouraged by the jointly managed social funds. Bad-weather allowances, vacation pay, and initial and ongoing training contribute to human capital formation and preservation in the German construction labor market.

However, labor standards, collective bargaining, and human capital formation have all come under greater competitive pressure in Germany compared to the Netherlands or Denmark. After the unification of East and West Germany, the less productive East German construction industry could not compete with West German contractors on the basis of quality and technology or on the basis of a skilled workforce. The regulatory framework in West Germany was applied to East Germany but not enforced. In addition, German construction began to employ posted workers from Eastern Europe and other European Union (EU) countries. Minimum wages for these workers, while required, are not generally paid in practice. Illegal practices in construction have become widespread.

Therefore, on the one hand, Germany has a highly productive and technically advanced construction industry with a highly qualified workforce. On the other hand, Germany has a rising proportion of low-skilled, often illegal workers using lagging technology and limited capital equipment. As a result, Germany has the beginnings of a two-tiered construction industry.

In response to this development, new regulations have been proposed, including a chain responsibility act similar to the one in the Netherlands, which would hold the general contractor responsible for the labor practices of subcontractors. Also, the Germans are considering prevailing wage regulations similar to those found in the United States. Whether Germany will continue to slide towards a two-tiered construction industry structure or whether new regulations will reverse this trend remains to be seen.

Denmark

Nikolaj Lubanski shows how the Danish construction industry also keeps free-rider strategies at bay. As in the Dutch case, in Denmark this is done through regulation, but the regulations are largely the result of collective bargaining with little direct involvement by the State. Key is the fact that 90–95 per cent of construction workers in Denmark are unionized. Moreover, posted workers and contractors also typically operate under collectively bargained agreements. Therefore, the provisions of collective bargaining, in effect, regulate the entire construction industry in Denmark.

Specialized educational initiatives for Danish construction workers are funded by levies established in the collective agreements. Posted workers are paid according to the wage rates agreed upon through bargaining. Additional unemployment insurance benefits do not emerge from collective bargaining simply because state-provided unemployment insurance economy-wide guarantees 90 per cent of the former income of an unemployed worker. Through collective bargaining, contractors must pay for quality insurance that guarantees to the customer the quality of construction for many years after it has been completed.

In recent years, the social partners in Denmark have started efforts to reduce or abolish demarcations between construction trades and to promote new types of work organization. There are also joint efforts to implement innovations to make small and medium-sized Danish construction firms more competitive, especially with respect to large Swedish companies that are bidding on projects in Denmark. Of all the cases in this study, Denmark has the formally best qualified construction workforce, and it is highly productive.

Canada

In Canada, where labor market regulations are decentralized down to the provincial level, Jean Charest presents the province of Quebec as a distinctive case both for Canada and in contrast to the United States. In the North American context, Quebec construction has a very high unionization rate of 47 per cent. Union membership for workers and contractor membership in employers' associations are legally mandatory in some subsectors of the industry. Collectively bargained agreements are generally binding for all workers and contractors within the construction industry whether or not they are members of corresponding unions or employers' associations. Mandatory affiliation is supervised by a joint labor-management body, the Quebec Construction Commission (*Commission de la Construction du Québec*), funded by a 1.5 per cent levy on the wage bill of each employer. In addition, other contributions are paid to funds for pensions, vacation pay, and training. The Commission operates these funds and, in effect, assumes some of the roles of employer by paying workers their vacation pay and pensions. Young workers receive initial training from provincial government-financed programs. The Commission finances ongoing training. It also certifies the competence of construction workers; all workers in the industry must hold a certification of competence from the Commission.

Quebec is an exceptionally regulated case in the North American context. It is supported by the actors within the Quebec construction industry, but it is under both political and economic pressure from outside construction. Economic pressure comes from a difficulty in recruiting younger workers into this industry. Given Quebec's severe winter climate, the average annual total of work hours for most construction workers is unusually small, 857 hours per year. Furthermore, 40 per cent of Quebec construction workers do not qualify for unemployment benefits because they have not accrued a sufficient number of hours worked per year. Thus, while the Quebec construction industry seeks to create a culture of training and to develop down a high-skill growth path, the limited number of hours of work on offer in this industry has limited the attractiveness of this career path for youth. Political pressure comes from broader pressures towards deregulation found outside construction, outside Quebec within Canada, and in the neighboring United States.

Australia

Economy-wide deregulation in Australia occurred throughout the 1990s, but strong unionization in the construction industry has withstood most of the effects of this overall deregulation. The Australian construction labor market used to be highly regulated. Elsa Underhill points out that trade union density and trade union militancy was high in this industry. Minimum standards that could not be undercut were guaranteed by a complex web of awards and agreements. An extensive array of portable, industry-based funds was collectively negotiated. This enabled employees to move to other construction companies without losing their vacation, sick leave, and pension entitlements. In addition, training and industry-specific research were supported by the industry funds, which were financed through employer contributions that amounted to 16 per cent of weekly earnings. These standards were enforced through a potent site-level presence of unions covering all firms including the subcontractors. Only subcontractors covered by pattern agreements could be hired by a principal contractor. The tendering process required compliance with these standards and agreements and there were high penalties for subcontractors not following the agreement.

This was the status of Australian construction when economy-wide deregulation was introduced. In Australia's overall economy, national collective bargaining has been replaced, since 1991, by decentralized enterprise bargaining as the primary method of determining wages and employment conditions. Labor market deregulation in Australia has been aimed at weakening wage arbitration decisions in favor of more variable, market-driven enterprise agreements. However, thus far, in construction, outside of residential work, strong unions and centralized collective bargaining have withstood this trend. Also, economy-wide, deregulation has been associated with increased job insecurity and an increased casualization of the labor markets. However, Australian construction has always been characterized by casual employment relationships and had built up a regulatory framework in the context of casual firm–worker relations. Finally, economy-wide, deregulation has been associated with increased wage inequality, as less unionized workers and those with little individual bargaining power experience relative declining wages. Despite the continuance of centralized collective bargaining in construction, the industry as a whole is sharing in this particular nationwide trend towards wage inequality. Deregulation in construction, however, has taken place mainly through structural changes in the industry. The percentage of self-employed who were never covered by the awards and agreements has increased substantially. Tax incentives have made self-employment attractive. The increasingly smaller size and the higher specialization of subcontractors have produced disincentives for training. Whether construction unions and centralized collective bargaining can withstand broader national trends towards deregulation and construction industry restructuring in a future of possibly slower growth or recession remains to be seen.

Spain

Job tenure was one way the Franco dictatorship in Spain sought to win public support. In the post-Franco era, after 1975, the government opened the economy to international competition, aligned its rules to EU standards, and extended the legal possibilities for employers to sign temporary labor contracts with workers. Temporary contracts have subsequently become common in Spanish construction. Justin Byrne and Marc van der Meer show that more than 60 per cent of Spanish construction workers are now under temporary contracts and have low firm tenure. In addition, the percentage of self-employed and illegally employed workers has increased. In the overall economy, Spain has a widespread informal sector that reflects the incapacity of the State to enforce regulations. This general tendency is more pronounced in the construction industry, and the weakness of unions in Spanish construction contributes to the inability of the State to regulate this industry. In contrast to many countries where construction unionization is higher in construction compared to other sectors of the economy, trade union density in Spanish construction is lower than the overall economy (11 per cent compared to 18 per cent). Employers' associations are also weak in Spanish construction. Surprisingly, in spite of this, due to trade union maneuvering in a booming industry under socialist government rule, collective bargaining in construction occurs since 1992 on a national level. Furthermore, the provisions of these contracts are extended to all construction workers through state regulation. In addition, the social partners have set up a national Labor Foundation for the Construction Industry (*Fundación Laboral de la Construcción*, or FLC), which is entrusted to provide vocational training and to promote craftsmanship, health and safety. This Foundation is funded by a mandatory but small levy on annual salary.

In many ways, Spain presents a less well-developed and less well-implemented version of the Quebec case. The FLC has funded some training and provided some studies on health and safety, but it has not succeeded in playing an active role in accident prevention. The collective agreements that were extended by the government to non-union members and non-affiliated enterprises are not implemented by many companies. Furthermore, self-employed workers are not covered by these extensions, and implicitly illegal workers do not receive the benefits of these contracts. On balance, in Spain, the presence of an industry-wide training fund, national collective bargaining, and an extension of bargained wages and conditions to non-union workers have been offset by the proliferation of temporary contracts, uncovered self-employment, and illegal workers. The inability – and probably the unwillingness – of the Spanish government to implement existing regulations have helped tip the balance towards a general state of deregulation in Spanish construction.

United States

Peter Philips shows that the United States has a two-tiered construction industry corresponding to a dual regulatory environment. The essential regulation governing construction in the United States is the prevailing wage law. This law requires that, on public construction, workers be paid the prevailing wage rate for an occupation in an area. Where union density for a construction occupation in an area is high, the prevailing wage rate is the collectively bargained rate. In areas where union density is low, the prevailing wage rate is the average wage rate for a construction occupation. Prevailing wage laws are more likely to be enacted in states or municipalities where union density is higher, and the presence of prevailing wage laws promotes the practice of collective bargaining or at least stems some of the forces that undercut collective bargaining in an area. Collective bargaining varies widely in the United States, with, for example, over 50 per cent of the Illinois construction workers being unionized while less than 2 per cent of the North Carolina construction workers are unionized. Currently, 31 states have prevailing wage regulations while 19 states do not. Where public construction involves federal funds, a federal prevailing wage law applies regardless of whether there is a state prevailing wage law in effect in that area. Prevailing wage regulations do not cover private construction, but many private contractors, particularly in prevailing wage law states, negotiate through collective bargaining. Collective bargaining provides through contract the regulations needed to establish funds for training, health insurance, and pension benefits in the United States.[2] Union workers organized along traditional craft lines move from union contractor to union contractor, and their pension and health benefits follow. Non-union workers typically get lower wages, fewer benefits, and no portability of benefits between contractors.

This regulatory structure has led to a two-tiered construction industry with the non-union contractors doing considerably less training and relying on a lower-wage, less skilled and less experienced workforce compared to union contractors. Certain regions of the country such as the South and certain sectors of the industry such as residential construction are dominated by non-union contractors using a low-wage, less trained labor force. Other regions of the country, including notably the Northeast, Upper Midwest, and Pacific Coast, and other sectors of construction, including notably heavy industrial and high-way work, are dominated by a high-wage, better trained, unionized labor force.

The 1980s in particular were a period of deregulation where nine states repealed their prevailing wage law. The United States presents a case where the effects of deregulation can be viewed both over time and by comparison of different states. The main lesson of this comparison is that prevailing wage regulations and collective bargaining provide the structure under which training can take place, workers can be attached to the industry, and the dynamic efficiencies of a technically developing industry can be encouraged. Where prevailing wage regulations have been abolished, the practice of collective bargaining has declined and with it training and career attachment to the industry.

United Kingdom

The UK construction labor market used to be strongly unionized and regulated. Prior to the 1980s, the State was both an employer and an important customer of the industry. Of unique interest, the British government followed a longstanding policy of steadily purchasing public housing units from the construction industry. This gave a stability to demand that is generally absent from construction. This government policy was replaced by a policy of promoting individual home ownership. With the shift to private residential construction, the industry has been exposed to cyclical highs and lows that were previously dampened by state policy. In addition, the State introduced a variety of policies that encouraged the proliferation of self-employment in construction through the replacement of worker status, with its attendant legal protections and costs, by independent contractor status, which lacks these protections.

Mark Harvey analyzes the consequence of these policy changes. Currently, about 50 per cent of British construction workers are self-employed. This has created a two-tiered labor market, with the hourly cost of the self-employed being 20–30 per cent lower than the comparable cost of a wage worker. Wage workers bring with them the overhead costs of unemployment insurance, workers' compensation, and other insurance protections that the self-employed do not enjoy. With the advent of this two-tiered system, overall job tenure with particular employers and within the industry has been reduced substantially. Collective bargaining also has declined precipitously. Safety has been imperiled and a culture of tax evasion by all parties has proliferated. The cost of this tax evasion in terms of state revenues is substantial. The increase in self-employment has led to the "hollowed-out firm," which does little actual work itself and rather attempts to develop and manage subcontracting relationships aimed at doing the work. The work, in turn, flows downward through an articulated and fragmented subcontracting structure. Harvey distinguishes two types of subcontracting, one of which delivers flexibility without fragmentation (a traditional sort of subcontracting) and an emergent form that yields flexibility with fragmentation. Fragmentation consists of parceling out similar jobs to almost identical subcontractors, who then are encouraged to compete with each other on the job site for further work. Thus, subcontractors do not form a team of cooperative specialists but rather a collection of competitors through the work process. This raises issues of cooperation, coordination, and control. Formal training, once a hallmark of British construction, has declined substantially, and the qualifications of individual workers are increasingly difficult to ascertain. The UK provides a cautionary tale of the pitfalls of deregulation.

Korea

Historically, the essential regulation in the Korean construction industry has been the licensing of contractors. This regulation created high barriers of entry into the industry for start-up companies and slowed the movement of companies up the ladder of industrial structure within construction. For example, the general contracting license limited the number of large-scale construction companies within the industry. Jin Ho Yoon and Byung-Goo Kang show that the result of these product market regulations was the development of an oligopolistic structure of large companies with a multi-layer system of subcontracting flowing down through four or five layers of subcontractors. In contrast to this strong regulation of the product market, the Korean labor market was weakly regulated. Unions were weak and suppressed by the government. Few laws existed to protect workers. Nonetheless, the industry profitability provided by its industrial structure allowed for the payment of above-average wages to construction workers compared to Korean manufacturing.

With the abolition of these product market regulations in the late 1990s, price competition increased and working conditions deteriorated. Subcontracting was extended and the percentage of daily and temporary workers within the industry increased to almost 60 per cent. Most temporary workers are excluded from state-sponsored social security. Wages now are close to those in manufacturing, with any premium for the insecurity of construction employment now lost. To stem this deterioration in construction conditions, the Fair Trade Act on Subcontracting was enacted, prohibiting temporary labor agencies from working in the construction industry. This act has not been actively monitored nor enforced, and illegal activity is widespread. Construction productivity in Korea is losing ground relative to manufacturing productivity, and Korean construction productivity ranks poorly in international comparisons. High labor turnover and the multi-layered structure of subcontracting forestall training and other capital investment.

Korea presents the case of an economy where the overall construction labor market has never been substantially regulated, either through state regulations or collective bargaining. The regulations that did influence this market came through the regulation of industrial organization. Deregulation came via the same route, opening up access to this industry, and state efforts to redress some of the negative aspects of this deregulation have not been enforced.

SUMMARY

The construction industry literally builds the foundation of every economy. The turbulence of construction work makes this important task of the construction industry difficult indeed. The result across the nine countries under study has been the emergence of two types of construction industries. Some countries have sought to meet this challenge through collective bargaining, state regulations, and policies designed to make training possible, create careers out of casual labor markets, facilitate investment in human capital, and impose quality standards on workers and contractors alike. These countries have created an industry-specific labor market in which construction workers develop ties to the industry and construction firms can rely on a stable workforce. Under regulation and collective bargaining, construction has developed along a high-skill, high-wage growth path. This "high road" strategy faces the challenge of continuing to adapt existing regulations and collective bargaining practices to new economic realities.

Other countries have sought to meet the challenge of building the foundation of a modern economy without regulation or through deregulation of the construction industry and the promotion of price and wage competition. Without regulation and collective bargaining, the construction industry in these countries has developed along a low-skill, low-wage path. This "low road" strategy must confront the challenges of an articulating and increasingly fragmented subcontracting system, the proliferation of free-rider strategies, and the high turnover of an untrained and inexperienced workforce. In these countries, the construction industry relies on the use of cheap labor and does not invest extensively in new technologies or more effective forms of work organization or worker training.

The country studies show that labor markets, industrial relations, and productivity differ substantially for these two types of construction industries. In some countries – for example, Spain, the United Kingdom, and Korea – large numbers of construction workers are temporary or self-employed and bear all the risks of the market. In other countries – such as Denmark, the Netherlands, Germany, the Canadian province of Quebec, and Australia – most workers have a permanent job, or the risks of fluctuating demand are buffered by an extended industry-specific safety net. To build up and continuously modernize this safety net and the training of the workers in these countries, trade union density in the construction industry is higher than in the rest of economy and industrial relations are based on mutual trust among labor organizations, employers' organizations, and the State. The reason for this mutual trust are the common interests in building up and retaining a skilled workforce in the sector, with unions interested in protecting workers from job instability and employers interested in safeguarding their investment in training. In the more deregulated economies, the construction industry unions and employers' associations are weaker than in the rest of the economy. Because of the dominance of small and medium-sized companies and the high job instability and turnover of the labor force, the formation of an interest representation for workers and employers is extremely difficult. The result is lower productivity and poorer product quality.

It remains to be seen which strategy proves more successful in laying the foundation economy-wide for economic development. We think that, in the long run, highly developed countries cannot afford a construction industry in which quality, productivity, and innovation lag behind the levels of other segments of the economy.

NOTES

1 The regulation of labor standards of posted or seconded workers is a EU phenomenon. Posted workers are employed by a company of the home country but posted to work in a second, host country – a subcontracting across borders that was included as part of the "four freedoms" of European integration. In many European countries, labor agreements cover all construction workers working in the country, with the national standards extended to posted workers in the country via a territorial principle. In other countries, posted workers are treated according to the standards of their home countries via the principle of origin. This allowed for the emergence of two different contractual labor standards on a single construction site. National regulations based on the territorial principle came into conflict with the principle of origin based on the freedom to provide services. The member states of the EU could not agree to extend the territorial principle to all posted workers in Europe. Some countries like Portugal and Spain wanted to post cheap labor to more highly developed countries. The United Kingdom opposed any regulation. Other countries like Germany, the Netherlands, and Denmark wanted to extend national regulations to all workers in their countries. In 1996, the member states of the EU agreed on the "Directive on the posting of workers" (directive 96/71/EC of 16 December 1996), which left it to the home countries to regulate hours, pay, and health and safety for posted workers until 1999. To prevent posted workers from undercutting national standards, the Netherlands and Germany modernized their regulations to require a minimum wage for posted workers, and Denmark required that posted workers be paid the same rate as corresponding local workers.

2 In contrast to most of the countries in this survey, the US government provides relatively limited health and pension benefits through a system of taxation and government services. Health insurance, in particular, is primarily provided, if at all, through privately funded employer-provided programs. These health insurance benefits are difficult to provide through employers in the context of construction's high labor turnover except through multi-employer programs collectively bargained with a construction union.

Chaos in the industry caused by dereg which in turn takes the low road vs high road global competition pushes the industry in the UK US + Canada towards a hybrid system. As deregulation grows the cycle of skilled labour shortages during booming economies leading to inflationary pressures + eventually succumbing to a downturn will persist. Developed countries will suffer overall if they deregulate too far.

2 The Netherlands

Rules in revision: high-quality production in the Dutch construction industry

Marc van der Meer

One of the most regulated of all the construction industries in this study, the Netherlands presents the case of a construction industry that has avoided the pitfalls of free-riding competitive tactics, as in the US case; fragmented and chaotic subcontracting, as in the UK case; and an inexperienced, unskilled, and unsafe workforce, as in the Korean case. Yet success brings with it challenges. The key challenge of the Dutch case is to fine-tune regulations continually in the face of the new technological and economic realities in Europe.

INTRODUCTION

The Dutch construction industry exhibits a combination of continuity and change. The industry's historical imprint has created its current profile, which includes a prominent role for productive craftsmanship and vocational training, the dominance of small and medium-sized enterprises, a high level of self-organization and self-regulation of the product market and labor market in the industry through the partnership of employers and employees, the close relationship of the industry with the State, and the relatively hard physical labor conditions. In comparison with other countries, the Dutch construction industry performs quite well, due to a number of labor market institutions and practices, such as industry-wide collective bargaining, training institutions, and health and safety regulations.

In the Netherlands, employers' and employees' organizations have become strongly involved in the management, administration, and implementation of labor market policies and social security provisions of the corporatist type of welfare state (Esping Andersen 1996; Hemerijck, van der Meer and Visser 2000) on both the national and industry level. The construction industry is a prominent example of such social partnership (van der Meer 1998). Construction firms are relatively well-organized in local and sectoral interest associations, whereas workers take part in inclusive trade union organizations. Since 1894, when the first collective agreement was signed, employers' associations and trade unions have created a highly developed system of regulations regarding wages and labor issues. The social partners work together to retain and produce a sufficient, broadly qualified labor force bound to the industry despite the cyclical and flexible nature of the production process.

Since 1926, trade unions and employers' organizations have established in collective negotiations several bipartite (i.e., jointly managed) social funds for additional unemployment benefits, vacation pay, a bad-weather allowance, vocational training, early retirement, health and safety, industry-wide occupational pensions, and general policy-making. These institutions of self-governance have resulted in several collective benefits. First, they produce a qualified labor force with high productivity; second, they improve working conditions; third, they promote labor peace within the work process since wage-setting takes place outside the enterprise; and fourth, they enable information exchange between interest organizations and the government about external developments.

In recent years, this structure has been challenged, due to both a re-evaluation of sectoral, national, and international policy processes in general and the trend toward "flexibilization" in the production process of the industry in particular. Consequently, formerly sector-wide regulation has started to differentiate at the branch and work floor level, and industrial relations are moving toward a wider range of individual choice within a collective standard.

The main argument in this chapter is that the social partners in the Dutch construction industry are strongly interdependent, through bipartite organizations at the industry level, and that their clear understanding of their joint interest enables them to adapt to internal and external developments and preserve the basic characteristics of the "high road" of quality production. They are usually able to forge agreement, often after lengthy discussions, to modify existing regulatory mechanisms to accommodate both internal and external changes; this becomes clear as we examine recent developments, including the initiatives to extend collective bargaining to include more qualitative labor market issues and to apply negotiated agreements to both temporary agency workers and workers posted or seconded from other countries. The future road, however, is uncertain, and depends upon the power balance and ongoing relationships among interest organizations in the industry.

The chapter starts with a discussion of the basic economic characteristics in the second section, followed by an analysis of the regulation of the product and labor markets in the third section. In the fourth and fifth sections, the training system and the characteristics of the labor market will be described. In the sixth section, the main challenges to labor market regulation are outlined, and the final section presents conclusions.

ECONOMIC CHARACTERISTICS

The construction industry is important to the Dutch economy in terms of production and employment. The industry covers different branches of activity.

Normally a distinction is made between the production, maintenance and repair of houses, dwellings, offices and utilities (*Bouw- en utiliteitsbouw*) and the civil engineering of earth, roads, and water infrastructure (*Grond, weg- en waterbouw*). In addition, there are related branches, such as the finishing industry, the installation industry, the supply industry, and architects and advisory consultants. Each branch has its own interest association and forms of specific regulation.

Output in the construction industry is seen as an early indicator for the Dutch business cycle, though there is variance in economic performance between subsectors, and the division of labor among companies changes over the years. In 1998, the gross value added of the construction industry was 5.36 per cent of gross domestic product (GDP); construction employment was 6.22 per cent of total employment. In Figure 2.1, the development of the economic conjuncture for the period 1969–2000 is shown, charting the substantial cyclical fluctuations of the industry and the rise of productivity.

In the period of recovery from World War II to 1970, the construction industry showed uninterrupted economic expansion in all categories and industries. After 1970, production and employment levels stagnated and even declined, reaching their lowest point in 1983. This resulted in a so-called "cold restructuring" of the industry by way of bankruptcy and job loss. The industry also had to redefine its position vis-à-vis its most important customer, the government, which moved to restrict its Keynesian government demand for construction services.

Since the late 1980s, the construction industry recovered in a stepwise fashion, though the level of employment lagged behind the growth in production. Arguments for the subsequent increase in labor productivity include the innovation of building materials, the automation in offices, increased prefabrica-

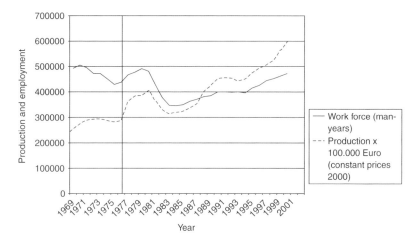

Figure 2.1 Production and employment, 1969–2000.
Source: CBS 2000.

tion, and increased flexibility in the use of workers and machines resulting in a better organization of the building process itself (EIB 1994).

After a new economic dip in 1991–3, all segments of the construction industry moved forward in the period 1994–2000. Output levels have increased annually; the output level of €57 billion in 2000 was produced by 472,000 persons working in the industry (CBS 2000). Though at mid-term a moderate decline in employment levels is expected of 0.2 per cent per year, the demand for construction workers stays high, with many signs of scarcity and vacancies, especially due to the large annual flow of workers out of the industry of 12 per cent per year (EIB 1999a).

Polarization of firms

Construction firms in the Netherlands vary in size and activities. Despite the existence of a few very large construction companies operating internationally, it is small and medium-sized enterprises that are responsible for the largest share in employment and production in the country. The number of companies in the construction industry proper has risen from 41,880 in 1994 to 62,090 in 1999. Approximately 99 per cent of firms have fewer than 100 employees, with most (54,185) having fewer than 10 employees. Of these, 28,925 are self-employed workers, a fast-growing category. There are 7490 medium-sized companies with 10–100 employees and only 415 companies with more than 100 employees (AVBB 2000).

In comparison with the smaller enterprises, the larger enterprises offer a total product by taking responsibility for all aspects of the architecture, design, planning, management, structural work, and finishing activities. Increasingly, the larger contractors have taken on a more coordinating role within the building process, handling negotiations with clients, architects, and government agencies, while contracting out the actual work to contracting partners and subcontractors.

The 400-odd larger enterprises are striving to diversify their activities to reduce the risks associated with a one-sided portfolio. Through mergers, cooperative agreements, strategic partnerships, and other economies of scale, they integrate technical know-how and economic resources. Some main examples of such larger companies are Volker–Wessels–Stevin, BAM–NBM, and Dura–Vermeer. Many times, mergers are set up to achieve economies of scale and to increase shareholder value, while in practice the formerly independent companies keep their own customers and manage their own projects at the local level.

The construction industry in the Netherlands used to be a solely national market, sheltered from international competition, with the exception of the water management and dredging companies, such as Boskalis and Ballast Nedam, which each maintain a substantial fleet of ships. Since the increase in European integration, more companies have taken on work within neighboring countries, especially Germany and Belgium, but the economic impact of this cross-border work is moderate; for example, in 1994 export accounted for less than 3 per cent of total output. The largest Dutch construction company, HBG, ranks only about 25th among the top European companies.

Table 2.1 Share in gross turnover of construction, by percentage of contracting and subcontracting firms

Year	Gross turnover (in millions DFL)	Percentage of contractors	Percentage of subcontractors	Others (%)
1976	31,708	82.7	11.6	5.7
1982	38,350	82.9	12.8	4.3
1988	52,109	81.9	14.7	3.4
1994	65,068	78.2	18.7	3.1
1997	74,765	76.7	20.6	2.7

Source: EIB 1999b.

Subcontracting

Because the larger companies are adopting more of a role as general contractors, subcontracting is an increasing, though still moderate, phenomenon. In the last 20 years, the share of gross turnover for principal contractors declined by about 6 per cent, whereas the share for subcontractors almost doubled (see Table 2.1). The growing use of subcontractors is due to several factors. First, the discontinuity of economic investment motivates principal contractors to spread risks. Second, they can take advantage through subcontracting of the increased degree of specialization within certain trades (e.g., bricklayers, road workers) that is due to technological innovation and changes in the work process. Third, various social security fund legislative measures have made employers reluctant to directly hire – and take responsibility for – operative laborers. Instead, they place the responsibility indirectly with subcontractors (EIB 1997).[1]

Subsequently, the composition of the workforce of principal contractors is changing. While in 1986 contractors employed one administrative or technical person for every three craft workers, almost 10 years later this ratio had been reduced to one-and-a-half-to-one (Interview with employers' representative, VG-Bouw 1995). Simultaneously, an increasing proportion of the labor force is employed at the smaller, subcontracting firms. In 2000, a principal contractor employs on average 19 persons; a subcontractor, 9 persons. Within the workforce, 67 per cent of all employees work for a principal contractor, 25 per cent work with a subcontractor, and 8 per cent work for companies that serve at various times both as contractor and subcontractor (EIB 2000).

THE REGULATION OF PRODUCT AND LABOR MARKETS

The regulation of the product market

In the Dutch product market, there is no law governing the trades, in contrast to Germany's *Handwerksrecht*. The basic regulation is the Act on the Establishment of Firms (*Vestigingswet Bedrijven*), introduced in 1937 with the aim of channeling competition based only on cost in the direction of competition based on quality and innovation. The Act, which is not restricted to the construction industry, did this by requiring technical and managerial competencies to enter the market (van Waarden 1989). After lengthy discussions between the government and representatives from the industry, the product market was deregulated in 1996 and 2001. As a result, the requirements for establishing a construction enterprise nowadays are significantly less demanding. A manager needs to demonstrate only knowledge of appropriate measures for health and safety and environmental protection. In the words of one personnel manager, "You can put a sign on your door and start a firm." To raise managerial standards, the employers' associations in the industry have developed a training program for directors and supervisors (*Stichting Kader- en ondernemersopleiding*), with attention to general information and project management, and the successful completion of this program serves as a standard certification for contractors.

In contrast, however, technological requirements and prescriptions on the relationships between contractors and customers are sharply regulated by the government in the Construction Agreement of 1992 (*Bouwbesluit*). This includes strictly defined standards for the selection and use of materials, quality standards for output, and rules in case of defects in the execution and supervision of the building process by contracting firms. The government applied, after consultation with representatives of the industry, many of the standardized model contracts and specifications for output and performance that had been developed earlier by business associations in the industry. These criteria were defined in terms of measurable minimum requirements to be respected by contractors.

Similarly, the sector has a long tradition of developing production standards. Nowadays, this development is done by the private Netherlands Standardization Institute (*Nederlands Normalisatie-instituut*, or NEN). This institute develops standards regarding the safety of buildings and utilities, environmental issues, communication in the building process, and attestation criteria for building materials and final products. Recently, firms have also begun to distinguish themselves and improve their competitive position through certification to various quality management standards, such as ISO 9000 and security and safety guarantees. With regard to environmental regulations, the industry has established a joint National Policy Plan for the Environment. This policy plan deals with issues such as recycling of construction and demolition waste, environmental policies within firms, education and publication, research and development, water cleaning, and environment-friendly techniques and products. This plan comes on top of national legislation for housing, building materials, and the environment, which are all serious forms of administrative regulation with legal sanctions and penalties in the event of misuse.

An increasingly important, but relatively recent phenomenon, is the growth in European regulation. Specifically, the Single European Act created the European internal market after 1992, and subsequent regulations have had a lasting effect on Dutch industry. These include the posted workers directive (1996), the recognition of qualification degrees (1992), and the recommendation for public procurement (1993). European regulation has increased the already fierce level of competition within the industry. For example, within the Dutch formal institution of the Price Regulating Organization (PRO), construction firms that are members of an interest association were bound by a Code of Honor of Building Contractors. This Dutch code required construction companies to report any intended bid; moreover, the winner in the bidding process was required to compensate the other bidders for their bidding costs. This form of price determination was allowed by the State to prevent pitting contractors against one another (Unger and van Waarden 1993: 63–4). However, the European Court of Justice declared this practice to be a form of cartelization and therefore illegal; appeals have been rejected (European Court of Justice 1996). In 2000, again after extended consultation between the Dutch government and business associations, a new Uniform Regulation for Tendering (*Uniform Aanbestedingsreglement*, or UAR) was enacted, to reduce transaction costs and create the option for sharing the costs of bidding.[2]

Overall, the regulations of the product market are weak regarding the constraints on market entry but strong regarding construction contract law. The legal standards reduce the potential for enterprises opportunistically to raise benefits at the expense of high-quality production. In recent years, a process of deregulation has taken place, though the social partners attempt to fill in the gaps through self-regulation of quality standards. This is only possible due to the strong inclusive membership patterns of interest associations, as will be seen in the next section.

The regulation of the labor market

Social partners

Companies and employees in the construction industry are relatively well organized. The trade union density for the industry is 40 per cent, compared to the national average of 28 per cent. Approximately one-quarter of union members are "inactive" or "non-working." The high trade union membership rate is the consequence of an institutional characteristic in the industry, where the trade unions until 1997 maintained a personal delegate in the execution of social security programs at the local level. This allowed for intensive contact of the unions with employees, and therefore for continuous membership recruitment, comparable to the so-called Genth system of social security which has led in Belgium and Scandinavia to a high-level of trade union-membership. In 2000, however, the social partners lost the government administration of this system (see Table 2.2).

There are two main labor unions. The largest is the social democratic Dutch Federation of Trade Unions (*Federatie Nederlandse Vakbeweging*, or FNV) with 160,000 members, followed by the Christian Union Federation (*Christelijk Nationaal Vakverbond*, or CNV) with 51,000 members. The machine drivers' association, the Black Brigade (*Het Zwarte Corps*), with 9000 members,

Table 2.2 Overview of the main social funds for employees covered by the collective agreement for the construction industry

Fund	Year started	Notes
The social fund for the construction industry (*Stichting Sociaal Fonds Bouwnijverheid*, SFB)	1952	Administered the execution of national social security policies (Unemployment Act, Sickness Act and Disability Act) for the whole industry, but lost this responsibility in 2000. It still administers as an umbrella organization the sectoral social funds.
The vacation and risk fund (*Vacantie- en Risicofonds voor de Bouwnijverheid*)	1947	Accumulates the contributions for wages during vacations by employees throughout the working year. After World War II this fund was combined with the "risk" or "bad weather" fund, which compensates enterprises the loss of working days due to frost and rain. After 1997 several "own risk" clauses have been introduced into this fund in order to reducing the high expenditures.
The pension fund for the construction industry (*Bedrijfspensioenfonds voor de Bouwnijverheid*)	1951	Provides for a sectoral pension. The contribution is at 15 per cent of the wage sum. Benefits are paid on top of the national general age pension.
The social fund for additional unemployment benefits (*Aanvullingsfonds WW*)	1988	Benefits are paid to unemployed construction workers during the first 8 weeks after unemployment, when the benefit level amounts to 80 per cent instead of the ordinary 70 per cent.
The social fund for labor conditions (*Stichting Arbouw*)	1985	Undertakes activities for the benefit of health and safety at work.
The social fund for active labor market policies (*Stichting Bouw-vak-werk*)	1989– 2000	Goal of this foundation was the improvement of allocation of supply of and demand for labor by providing statistical information, work experience programs, and labor market allocation. In 2000, it expired in the discussion on social security in the Netherlands.
The education and development fund (O&O-fonds, *Stichting Opleidings en ontwikkelingsfonds*)	1967	Compiles via a 0.43 per cent contribution on gross wages resources for, first, research and development, second, the sectoral training institutions, and third, a substantial financial contribution to the trade unions and employers associations, signing the collective agreement. All industry research institutes (such as the EIB, *Economisch Instituut Bouwnijverheid*, which is conducting surveys among employers and employees) are paid from this fund. Also the institutions for vocational training in building and utilities (*Stichting Vakopleiding Bouwbedrijf*, SVB), since 1946, and in civil engineering (*Stichting Beroepsopleiding Wegenbouw*, SBW), since 1943, are likewise financed. Finally, the employers' and employees' organizations signing

Table 2.2 (Continued)

Fund	Year started	Notes
		the collective agreement distribute €9 million on an annual basis for compensation of their internal staff.
The early retirement fund (*Stichting Uittreden Bouwbedrijf*, SUB)	1978	Based on the pay-as-you-go system until 1995. Since then a capital accumulation principle has been introduced. Nowadays people save for their own early retirement.
The social fund for days off in the work schedule (*Stichting Fonds Roostervrije Dagen*)	1985	Created after the introduction of the collective working time reduction in 1985. Each week, four out of 40 working hours are accumulated for additional holidays for workers in order to restrict overwork and increase the opportunities that the unemployed are being hired to a greater extent.
The training fund (*Stichting Scholingsfonds voor de Bouwnijverheid*)	1988	Initiates retraining for the employed incumbent worker. The contribution is at 0.8 per cent of the gross wage bill.

Source: van der Meer 1998.

cooperates with delegates from FNV and CNV to negotiate on behalf of construction workers.

There is a broad variety of employers' and trade associations, and all major firms are affiliated with one or more representative organizations, each with its own domain and influence. Van Waarden (1989) counted 220 different trade and employers' associations at the national and regional level for the construction industry, based on data from 1980. Nowadays, as the result of a process of amalgamation and scale enlargement, there are 30 distinct employers' associations for the various branches of building, utilities, civil engineering, rail infrastructure, dredging, and water management (SFB 1998). Within the industry, different umbrella federations exist, among which the Federation of Dutch Contractors Organizations (*Algemeen Verbond Bouwbedrijf*, or AVBB) is the most representative.

The AVBB organizes 18 associations in the construction industry. Among these, the most important are Dutch Association of Contractors in the Construction Industry (*Nederlandse Verbond van Ondernemers in de Bouwnijverheid*, or NVOB) for small and medium enterprises and the Association of Larger Construction Enterprises (*Vereniging Grootbedrijf Bouwnijverheid*, or VG-Bouw) for the larger enterprises. NVOB and VG-Bouw have formed a federation called *BouwNed* to reduce transaction costs and interlocking directorates. In December 2000, the Confederation for Subcontracting Firms (*Confederatie van Gespecialiseerde Aannemers*, or CONGA) affiliated with AVBB, so AVBB now also represents the larger subcontracting enterprises, where about 10 per cent of the labor force is directly employed.

Collective agreements

In the construction industry, there are 11 collective agreements in different branches. Perhaps, the most important is the collective agreement for manual craft workers in the construction industry, which covers 150,000 workers in both building and civil engineering. Supervisory, technical, and administrative staff personnel (40,000 persons) in building and civil engineering have their own collective agreement. Separate collective agreements also exist for painting, finishing, and glass workers; natural stone workers; wet-grind dredge personnel; mortar transport workers; plasterers; plumbers, fitting, and heating personnel; roofing workers; and dredge personnel. In the related wood industry, other collective agreements have been signed, such as furniture industry; wood trade; carpenter firms; wood and brush industry; packaging, pallet and wooden shoe industry; house design firms; yacht construction; and the corporations that lease housing.

Wages and employment conditions for the 11 branches of the industry are set through national collective bargaining. Trade unions and employers' associations annually (or sometimes biannually) forge agreements that since 1927 have been granted the status of law for their respective domains. Due to *erga omnes* (Latin: "in relation to everyone") clauses, these agreements are also binding on non-unionized workers employed in those enterprises that are affiliated with an employers' association. Moreover, since 1937, collective agreements have been routinely declared by the Minister of Social Affairs and Employment to be "generally extended" even further, to workers in unaffiliated enterprises. For example, in the construction agreement that covers manual craft workers, in addition to the 5000 enterprises affiliated with an employers' association, whose 120,000 employees are directly covered through *erga omnes* bargaining, the 8000 non-organized firms with 30,000 employees are also covered through general extension (Ministry of Social Affairs and Employment 1998).

The trade union organizations have a reputation for strength, whereas the employers' associations, whose members compete with each other in the product markets, have often been internally divided. In their wage policies, the trade unions uphold some normative principles, which include a cost-of-living allowance, a proportionally equal wage increase for all workers, no application of piece-rates, a ban on overwork, and respect for health and safety. It is due to their power that the collective agreement covers so many issues, and they are responsible for the collective social funds. However, there is one area considered to be outside the competence of the labor unions: in contrast to the restrictions of management decisions by demarcation of crafts in the United States, the allocation of labor within the enterprise is a prerogative reserved solely to management.

The wage schedules in the collective agreement prescribe minimum rates. At the firm level, employers and individual employees may negotiate additional salaries and bonuses. In the current tight labor market, 66 per cent of employees on the work floor receives a performance-related bonus, at a fixed level averaging 16 per cent (EIB 2000). Approximately 10–15 per cent of the labor force occasionally works with piece rates. In addition, the collective agreements regulate allowances for clothing, boots, and transportation. The work week is 36 hours, divided over 5 days of 8 hours, with 4 hours each week allocated to additional vacation and training leave.

Social funds

Through collective bargaining, a labor market and social security policy has been developed to manage the social funds for the sector, which are in addition to the national social security provisions. This process started with the establishment of the first vacation fund for stone laborers in 1926, which was designed to avoid workers being dismissed for the Christmas holiday period. Accordingly, individual risks have been collectively regulated on a bipartite basis – i.e., with the agreement of both social partners.

Similar regulations have been created for social security, disability coverage, bad-weather allowances, industry pensions, voluntary early retirement, research and development including vocational education and training, and health and safety. These funds are financed by enterprises and workers themselves, with the total balance amounting to €224 million in 2000 (SFB, *Annual Report* 2000). Many of these funds are organized at the level of particular branches of the industry (e.g., building, plastering, painting, mortar, and roofing branches). The social security systems were for many years implemented by an industrial insurance board, *Sociaal Fonds Bouwnijverheid* or SFB. The employers' associations and labor unions were shareholders of this private organization. In 2000, the social security tasks were transferred to public control, in the aftermath of a parliamentary inquiry on social security in 1993 (Hertogh and Peet 1999; Jansen, Jeurissen and Lenoir 2001).

The coverages of the collective funds, described in Table 2.2, are agreed upon in collective bargaining, along with the contributions required from employers and employees. In 1998, the total contribution for these collective funds, excluding pensions, was 14.9 per cent of the gross wage bill (van der Meer 1998: 204). Since the mid-1990s individual enterprises have been able to choose whether to participate in some of the funds. For example, contributions to the bad-weather allowance fund were obligatory until 1998; now enterprises may decide whether or not they want to cover the risk of bad weather. Similarly, the early retirement fund was restructured after 1995 from a collective pay-as-you-go system into a system with capital accumulation funding for individual employees. These changes have occurred despite the trade unions' push to continue obligatory contributions.

Strategic policy consultation

The construction trade unions, FNV and CNV, and the federation of employers' associations, AVBB, also have a platform for general policy consultation and strategic information exchange (*Bouwberaad*), comparable to the national bipartite Foundation of Labor (*Stichting van de Arbeid*), where the national confederations of employers and employees meet.[3] In the Bouwberaad, the social partners discuss socio-economic developments and share concerns, opinions, thoughts, and policy ideas, while building a database of relevant facts and figures. This results in a common base of knowledge on the socio-economic situation in the industry and extends the options for alternative policy positions and strategic choices. Consequently, the economic situation is no longer a point of discussion in collective bargaining, whereas alternative strategic policies have become a topic for negotiation. An example of a trade union initiative in the Bouwberaad is the establishment of the construction trade job (*Bouw-vak-werk*) foundation for the work experience of job seekers and the training of workers.[4]

At least twice a year the Bouwberaad consults with the Ministry of Housing, Planning, and Environment (*Volkshuisvesting, Ruimtelijke Ordening en Milieubeheer* or VROM) about governmental policy-making in the industry. Consultation also takes place in the Broad Administrative Consultation for the Construction Industry (*Breed Bestuurlijk Overleg Bouw*) with participation from the Ministries of Economic Affairs, Transport, Social Affairs, and Employment. In 1991, the government and interest associations in the construction industry established a European Council for the Construction Industry (*EG-Beraad voor de bouwnijverheid*), to discuss issues arising out of European integration. These platforms facilitate the consultation and negotiation on the implementation and adaptation of government regulation to the needs in the industry.

Additional regulation of the labor market

Compared to most other countries in this study, the labor market in the Netherlands is also strongly regulated concerning working conditions, working hours, and social security. The duration of weekly working hours has always been determined industry-wide through collective bargaining. The Working Hours Act of 1995 allowed the distribution of working hours to be determined by local management after consultation with the works councils at the enterprise level; in the construction industry, this is applied in some firms where works councils play a role. Regarding working conditions, in addition to the provisions of the Working Conditions Act of 1998, the representatives of the industry have concluded several initiatives regarding health and safety, including the joint fund *Arbouw* (see Table 2.2). Most recently, the industry has signed three "gentlemen's agreements" with the government to reduce the use of organic volatile compounds, to control work-related stress, and to improve the re-integration of disabled persons.

Furthermore, the employers' associations in the industry itself have also established a Foundation for Promotion of General Construction Interests (*Stichting Behartiging Algemene Bouwbelangen*, or SBAB) for controlling lawful entrepreneurship and tracking down illegal moonlighting, i.e., working without paying taxes and social security contributions.

In addition, in 1982 the Chain Responsibility Act (*Wet Ketenaansprakelijkheid*) was passed in the Dutch parliament to reduce illegal work. According to the Dutch model of reducing hazard in the production chain, the principal contractor is responsible for the payment of social and tax contributions in the case of negligence by the subcontractor. This enlists the help of the principal contractor in averting unreliable entrepreneurship. To protect the main contractor from misuse by subcontractors, a so-called "reserved account" (*G-Rekening*) was created, where tax and social contribution payments are saved during the construction process. Evaluation studies of this Act by the Dutch Social Security Council showed that fraud on tax payment declined and social security contributions were paid sooner, resulting in more peaceful market relations.

THE QUALIFICATION STRUCTURE AND TRAINING

In the guild structure of the medieval period, the master craftsman instructed his apprentices on the job. After the establishment of the first vocational and technical schools at the end of the nineteenth century, young people received their vocational education and developed their skills on the shop floor. Since the 1940s, trade unions and employers' associations have established their own sectoral training institutions for apprentices, one for building and utilities, established in 1946, and one for civil engineering, established in 1943. These national bodies are responsible for setting the standards for apprenticeship training, counseling training firms and apprentices, developing examinations, and rewarding diplomas. In 1967, the Education and Development Fund (*Opleidings- en Ontwikkelingsfonds*, commonly called *O&O fonds*) was established to spread and reduce the costs of training by a sector-wide levy on gross wages. This levy was extended by the government also to non-organized enterprises. In 1970, this fund, with the help of the government, paid 11 per cent of the training costs of apprentices; by the early 1990s, the contribution was 40–60 per cent (te Lintel Hekkert 1992).

From its start, this system operated relatively efficiently. However after the first oil crisis, the annual number of apprentices fell, from 11,000 in 1970 to 5000 in 1975 (te Lintel Hekkert 1992). Employers saw training as time-consuming and costly, especially in a period of economic crisis. When recession again hit the industry in the early 1980s and vocational training came under pressure once more, the local employers' associations created "cooperative associations" (*samenwerkingsverbanden*) to keep training going. These associations employ apprentices themselves and lend the apprentices to enterprises, with the financial help of the Education and Development Fund and a government subsidy.

In the mid 1990s, the cooperative associations raised other problems of the vocational training structure. They argued that the curriculum of the training programs was too general in nature, resulting in "apprentices who don't know to use a hammer." They pointed to the lack of a modular system, the high number of dropouts, and the continuous tension between government and the industry about who should bear the responsibility and cost of training of the unemployed. Training was suffering from the free-riding of smaller companies not affiliated with the employers' associations. Finally, there was not a smooth transition from initial vocational training to on-the-job experience.

Given these critiques, the State has gradually, though slowly, integrated the general secondary professional training and the apprenticeship scheme into one coordinated system of intermediate-level vocational education and training. Within this general framework, the industry organizations can develop the courses, which allow apprentices and incumbent workers to improve their official qualifications systematically. This reform process culminated in the 1997 Adult and Vocational Education Act. There are now 22 "new style" national bodies for training, including one in the construction industry. These bodies are responsible for the qualification structure, i.e., the set of skill standards for vocational education tracks, at four levels of qualification: (1) assistant; (2) basic skilled worker; (3) skilled worker; and (4) middle management or specialist. The second level is perceived as the required minimum level for craftsmanship, comparable to the German *Facharbeiter* or the American journeyworker. Level 4 corresponds to specialized craftspersons. At each of these levels the competency standards for all professions within the trade (carpenters, bricklayers, etc.) have been formulated.

The level of education of the construction workforce is lower than that of the Dutch labor market as a whole; however, it is significantly higher than that of construction workers in most of the other countries in this book. As can be seen in Figure 2.2, workers with only a primary general education – and therefore no formal qualifications for construction work – comprise 23 per cent of the building and utilities workforce and 27 per cent of the civil engineering workforce. With some lower general vocational training and a specialization in construction are 47 per cent of building workers, whereas 23 per cent have finished lower general vocational training with a specialization other than construction (in sum 70 per cent). For civil engineering these percentages are somewhat higher, 50 and 26 per cent respectively. Finally, 7 per cent of the workforce in both building and civil engineering has at least secondary general vocational training. In the figure, these percentages for workers in building and utilities add up to 100.

In addition, substantial progress is noted regarding industry-specific professional vocational training (at levels 1–4). The number of employees with a recognized certificate rose from 35 per cent in 1990 to 43 per cent in 2000. Though 57 per cent have no official industry-specific degree, 28 per cent have finished the primary craft worker training (levels 1–2), and 13 per cent have finished the secondary craft worker training (levels 3–4), whereas 2 per cent have a degree of the entrepreneurial management training.

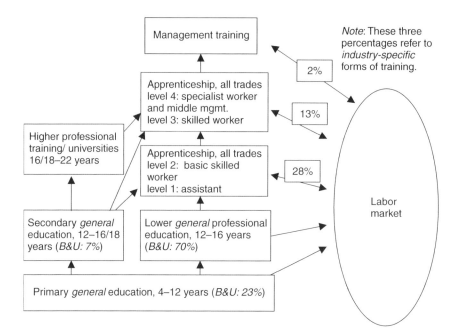

Figure 2.2 Professional education and training in construction (building and utilities only).

These days, "life-long learning" and "employability" are at the top of the agenda in the Netherlands as a joint instrument for employers and employees to anticipate eventual job loss by keeping skills up-to-date. The Dutch collective agreement for construction workers has contained an article (35.b) about the right to life-long learning since 1988. The costs for continuous training are paid by a 0.8 per cent premium on the wage bill into the sectoral training fund. The article stipulates that all employees have the right to receive two days of training per year, and the fund pays 100 per cent of course costs, approximately €10 for traveling expenses, and compensation for training hours. These training days may be used to prepare workers for other jobs in the future. Larger firms, especially those with a specialized human resources department, use the funds more extensively than smaller firms, who can hardly afford to lose a worker's labor for a few days. Nonetheless, in 1998 and 1999, an estimated 200,000 training days were provided for employed workers – double the rate in 1995 – and the average participant participated in 3 days of training per year (Scholingsfonds, various years).

Overall, training programs in the construction industry have developed substantially over the last two decades, becoming more sophisticated in terms of content across a broad range of fields pertinent to the success of the industry. The industry has gained control over the modular curriculum, within the context of the general governmental legislation. Moreover, thanks to the efforts of the employers' associations, apprentices are guaranteed opportunities for gaining work experience, while the industry collectively pays part of the costs. Finally, the industry facilitates life-long learning via the collective agreement, to the benefit of both employees and employers within the industry.

LABOR MARKET CHARACTERISTICS

In the 1970s and 1980s, operative construction workers often were hired on a temporary basis for the duration of a building project. Nowadays, long-term relationships between enterprises and employees are more typical, in spite of the fluctuating economic climate and the greater number of small companies. The majority of workers in the industry today are working full-time with permanent contracts. In 1997, 83 per cent of the workforce in construction had a permanent contract; only 1 per cent, a temporary contract; 16 per cent were self-employed (CBS 1999). This compared positively with the national labor market (see Table 2.3). This stability for construction workers is possible because the system of regulations reduces for individual employers the risks associated with providing steady employment. For example, wages are partially guaranteed through a spell of bad weather by means of the collectively financed bad-weather allowance. Moreover, when lay-offs do happen, they are more easily accepted by trade unions because of the extended unemployment compensation provided to laid-off workers.

Along with the investment in preserving the employer–employee relationship on a permanent basis has come an investment in skills and qualifications. Over time this has developed the labor force into a highly productive, treasured asset, which is not casually replaced by unemployed job seekers. The average tenure in a given enterprise is 10 years, and the average seniority in the industry is almost 18 years, whereas seniority of less than 1 year within one firm occurs for only 17 per cent of the employed working population (see Table 2.4).

Table 2.3 The employed labor force, by number of jobs, 1997

	Employee with permanent contract	Employee with fixed term contract	Total employees	Self-employed	Total employees and self-employed
Construction (including installation)	392,000 (83%)	5,000 (1%)	397,000 (84%)	74,000 (16%)	471,000 (100%)
Total economy	5,916,000 (74%)	722,000 (9%)	6,638,000 (84%)	1,308,000 (16%)	7,946,000 (100%)

Source: CBS 1999.

Table 2.4 Seniority in the enterprise

Length of service	Working with primary contractor (%)	Working with subcontractor (%)	Both (%)	Total (%)
<1 year	14	20	24	17
1–3 years	17	18	15	17
3–5 years	10	12	7	10
5–10 years	16	19	19	17
>10 years	43	31	35	39
Total	100	100	100	100

Source: EIB 2000.

Employers have realized that recruitment and skills training are costly tasks, and therefore long-term contracts are relatively efficient. For this reason, most companies maintain loyalty to an inner circle of permanent staff, and they create the additional staffing flexibility they need through such means as using over-time from their trained staff, subcontracting to other enterprises, and through "collegial hiring," where employers deploy the employees of another enterprise for a few days without officially hiring them. The allocation of employees often takes place directly between enterprises; therefore, mutual trust among com-panies plays a substantial role. Alternative forms of flexible recruitment – for example, through temporary labor agencies (forbidden from 1982 to 1997), fixed-term contracts for some employees, and public employment services – have been used to a far lesser extent (EIB 1998b).

Within companies, personnel management in the construction industry is fairly traditional. The relationships are predominantly top-down-oriented; i.e., managers typically instruct employees what to do, notwithstanding that relations are relatively informal with direct and open communication. Annual surveys among construction craft workers reveal that 70 per cent evaluate their working conditions as "good," while an increasing number judge their work to be highly competent, with room for improvisation, personal initiative, and decision-making (EIB 2000). The larger companies have more developed human resource management departments. Also, the larger firms are more likely to use institutionalized works councils, which have advisory, information, and negoti-ation rights.

The average age of the workforce is 38.5 years, and the workforce is aging. Younger people continue to enter the industry, but they often leave after a short working period. Labor conditions in competing industries are apparently more positive, although the level of rewards in the construction industry has increased 50 per cent over the last 20 years (EIB 2000).

The number of ethnic migrants (2 per cent) in the industry is very low compared to most of the countries in this study. Recruitment networks enable selection of especially male workers in mostly rural areas with relatives and families, who commute every day to workplaces in the cities. Ethnic minorities do not take part in these networks and are therefore represented on the shop floor to a lesser extent. When they enter the workforce via the training institutions, they face difficulties of being accepted as productive laborers. Indirect forms of discrimination dampen the chances for a long work history for these categories in the industry (van der Meer 1998).

CHALLENGES TO LABOR MARKET REGULATION

The social partners in the Dutch construction industry can take pride in their achievements. Through the cooperative development of regulations, they have built a modern, highly productive construction industry, capable of building the infrastructure of a technically advanced country. However, these achievements are put at risk if the regulations that permitted them cannot themselves be flexibly rebuilt to meet the continuing challenges of a dynamic world economy. In the following section, four main challenges to the labor market regulation are discussed.

Employment risks: the flip side of success

The current positive labor market situation obscures the employment difficulties of a decade ago when about half of the construction workers were not working. At that time, trade union advisor Kees Korevaar (1990) typified the construction industry as "extremely ill." The number of inactive construction workers of working age was as high as the number of actively employed workers. With a heavy reliance on the extended use of unemployment, disability, and early retirement benefits, the construction industry had begun to resemble the "welfare without work" setup of the corporatist welfare state (Esping Andersen 1996). In 1993, a parliamentary inquiry into social security in the Netherlands called the social partners "jointly guilty" in the misuse of collective social funds, which provided an exit mechanism from the labor market (Buurmeijer 1993).

Due to the subsequent privatization of the Sickness Act in 1996, sick leave became an individual risk for enterprises, which no longer could use the sickness regulation as an exit from the labor market. Subsequently, sick leave in the industry decreased by half during the 1990s, from 10 per cent on average in 1993 to approximately 5 per cent in the late 1990s. The inflows into the ranks of the disabled, however, have not seriously abated in spite of the individual-risk clauses that were introduced in 1994; currently there are 65,000 disabled construction workers, or 17 per cent of the total disabled workers in all industries. The elevated figures on disability are due to the number of accidents, which are still high in construction, though restricted in comparison to most other countries in this book. In 1998, 13,750 accidents occurred on the work floor, due to falls, heavy loads, stumbling, or sliding. In 73 per cent of the cases, one day of sick leave proved necessary (EIB 2000). The relatively high number of accidents has resulted in additional safety measures for scaffolding workers, road workers, and painters. In addition, new environmental measures are being taken.

Self-employed: neither employer nor employee

The rise in self-employment has worried both the trade unions and the employers' associations representing small and medium-sized firms. Compared to ordinary wage earners, the status of the self-employed is unclear. On the one hand, they are not employees, since they have no labor contracts, do not adhere to collective agreements, and do not contribute to the collective funds. On the other hand, neither are they employers, since they work under subcontracting conditions and have little control over the pace or substance of the work. In violation of the standards set by the collective agreements, the self-employed often work more than 40 hours per week. Moreover, they are insured on a voluntary basis only, and their levels of insurance against the risks of accidents, disability, and sickness are far less than that of wage earners who participate in the obligatory collective funds (EIB 1998a). In other words, the self-employed take individual risks by offering flexibility, lower prices (without social security costs), and longer working hours than employees covered by the collective agreement. For these reasons, the trade union FNV did not allow the self-employed to become members of their organization until 2000. The trade union CNV, however, still does not recruit self-employed persons.

"New" categories of employees: temporary agency workers and posted workers

Construction workers under the collective wage agreement in the Netherlands must compete with posted or seconded workers from other countries and with workers hired by temporary agencies. The social partners in construction have, through a series of steps, managed to include seconded workers and agency workers under the construction agreement.

In 1997, when the ban on temporary agencies was lifted, 6 per cent of all enterprises hired at least some temporary workers through such agencies, and approximately 40 per cent of enterprises considered this option to be positive (EIB 1998b). Agency workers have an advantage in wage costs, since the collectively agreed rates for these categories are lower, and since the agreements do not oblige firms to pay contributions to social funds. The social partners now require agency workers to respect the construction agreement in full (including the contributions to the social funds) once they have finished a degree in vocational training or have 12 months of work experience in the industry. This stipulation was exempted from the general extension of the collective wage agreement, and therefore it applies only in enterprises affiliated with employers' associations.

For seconded workers, the European Posted Workers Directive of 1996 allowed European member states to implement rules on the posting of workers from other countries at the national level (Eichhorst 1998). The Dutch social partners agreed in the collective agreement that seconded workers must respect the construction agreement after 1 month of work in the Netherlands. The social partners in the Dutch and Belgium construction industries agreed to respect the mutual stipulations for Belgium and Dutch construction workers even from the first day of work. Finally, the Dutch government enacted the 1999 Act on Employment Conditions for Seconded Workers (*Wet arbeidsvoorwaarden grensoverschrijdende arbeid*), prescribing that a generally extended collective agreement also covers seconded workers from the first day. The law refers, however, only to the stipulations within the collective agreement on minimum wages, maximum working hours, minimal rest hours, minimal vacation, conditions for posting of workers, health and safety, protection and equal treatment of employees. The additional industry-specific collective funds specified in the collective agreement are not included.

The need for "modernization" of the collective regulation

When the 1995 collective bargaining round for the construction industry resulted in a stalemate, the employers' associations initiated a reform process. The employers proposed to flexibilize labor hours further by not paying any more supplements for evening and night work and by allowing temporary labor agencies, which were at that time forbidden. Moreover, they suggested discontinuing the early retirement fund, the bad-weather allowance, and the vacation fund, arguing that these funds are inefficient and that enterprises can regulate themselves well.[5] The training institutions were exempted from this attack.

The spokesperson of the employers rhetorically commented on this: "Today's society is absolutely different from 30 years ago. There has been a shifting from the collective to the individual. In scaffolding, you shouldn't come around with a concept of solidarity." The strike that followed ("for solidarity" and "for a good collective wage agreement") resulted in the largest post-war labor conflict in the Netherlands, with 36,000 workers striking for 23 working days in Spring 1995.

That labor conflict was a critical episode in a period that saw both a change in the competitive relationships between firms and an adjustment in the systems for social security. The employers' association AVBB argued that in the extensiveness of the labor and product market regulations one can no longer "see the forest for the trees." It maintained that the collective agreements are too detailed and too expensive, and that costs should be borne by the branches that created them. Clearly, the employers' associations have an agenda in line with the "de-collectivization of legal measures," "the process of individualization in labor relations," and "the deregulation and the related flexibilization to overcome rigidity due to regulation" (AVBB 1996a; AVBB 1996b).

Since then, collective bargaining for the construction wage agreement in 1997, 1999, and 2001 has proved to be complex and full of tension. The parties have agreed to "modernize," via a jointly managed study, the formal structure of the collective agreement and the social funds (van der Meer 2001). They plan to establish a "framework" collective agreement covering several branches generally, with a specific agreement added for each branch. The general issues such as contract types, rights and obligations, working hours, and wages structure will be laid down in the framework agreement as far as they concern the industry as a whole. Tailor-made agreements referring to the functions and developments in specific branches as well as individual needs such as training, child care, labor conditions, and flexible opening hours, will be formulated in the separate appendices. Within this process, the effectiveness and efficiency of the social funds will be evaluated as well.

INSTITUTIONAL CHANGE AND FURTHER MODERNIZATION

The Dutch construction industry is a good example of a "social market economy," though its training and social policies differ significantly from those of Germany and the Scandinavian countries. The governance structure of the Dutch construction industry is highly institutionalized in several collective funds at the branch and industry levels. This produces a qualified labor force with high-quality output, due to both training and health and safety provisions; low levels of conflict on the shop floor; and opportunities for strategic decision-making between actors. The low-turnover profiles in firms and the reproduction of the workforce are especially worth mentioning, in spite of the flexible nature of the production process. Also the Chain Responsibility Act controlling the payment of tax and social security contributions, as well as the qualitative change of the collective wage agreement with its attention to life-long learning and individualized early retirement schemes based on capital accumulation, and the application of the collective wage agreement to posted workers and temporary agency workers, prove that adaptation of collective working rules to changing product markets is possible.

These adaptations have been possible due to the critical condition that associations of management and workers are broad, inclusive organizations that negotiate agreements that are actually implemented in the work process. In addition, the government plays a substantial role by setting a framework for implementing regulation, by extending the collective wage agreement generally to non-organized firms, and by controlling regulation in the enterprises.

In spite of these successes, the overall process of this form of collective regulation is put to the test of efficiency and effectiveness, due to several driving forces both at the enterprise level and at the policy level. At the firm level, new technologies have resulted in prefabrication, new products, faster production processes, and specialized cooperation between firms. Larger contractors increasingly take a coordinating role in the production of infrastructure and housing, leading to temporary coalitions of enterprises that are strong enough to develop a set of labor regulations independent of the industrial wage policy, while contracting out work to partners and subcontractors. Technology changes the level of craftsmanship of the workforce and leads to a need for more general management competencies and computer skills. The consequence is a process of desectorization, with less clearly defined boundaries between industries, and with the risk of crowding out relatively expensive construction labor in favor of cheaper collective agreements for wage earners or self-employment.

In addition, policy processes change. The internationalization of the rule-making machinery, especially the process of European integration, supports market forces and defines new criteria for nation-specific regulation. At the national level, social security and labor market policies will be further reorganized and integrated. This so-called "one counter" structure aims to achieve scale advantages and cost reductions through individual, rather than collective, options for social security. At the industry level, there is a change from centralized to decentralized decision-making on employment conditions, allowing for more individual solutions at the branch level and the shop floor level regarding the flexible application of the working hours, wages, and health and safety, within the structure of the collective wage agreement.

These developments do not imply that collective labor regulation will disappear, but they have resulted in a joint effort by the social partners to modernize the rules. The collective agreement will likely become leaner, more transparent, and more tailored to the needs of each branch. In these decision-making processes, the trade unions and employers' associations so far have adapted the existing regulations to external conditions. This shows an interdependence that is important for the reproduction of a broadly trained and productive labor force, needed for keeping the industry on the high road.

NOTES

1 Within the Dutch welfare state, many social security provisions are being reconsidered. In 1993, access to the Disability Act was restricted, and in 1994 "own risk" clauses were introduced for firms, placing the risk for employee disability payments with the employer. In 1996, the Sickness Act was privatized, making the risk of sickness an individual hazard.

2 After recent indications of fraud, the practice of tendering will be investigated by a special inquiry committee of the Dutch Parliament in 2002.

3 This Foundation of Labor is one of the core platforms of the Dutch consultation economy (Hemerijck, van der Meer and Visser 2000).

4 This sectoral initiative was created in 1989, after 5 years of discussion to find an answer to overcome the market and government failure of large unemployment in the 1980s. The exploration of strategic alternatives was initiated in the Bouwberaad, and the final decision-making was taken in collective bargaining. In the current boom period, the Bouw-vak-werk initiative has been expired in spite of substantial success (van der Meer 1999).

5 The "wedge," indicating the difference between gross labor costs (including social taxes) and net salaries, comprised 32.2 per cent for an employer and 12.9 per cent for

an employee in 1998. In this percentage, the 14.9 per cent contribution for the sectoral collective funds has been included (van der Meer 1998: 204).

REFERENCES

Algemeen Verbond Bouwbedrijf (AVBB) (various years) *Annual Review*, The Hague: AVBB.

Algemeen Verbond Bouwbedrijf (1996a) *Wetten en regels in de bouwnijverheid: van heden naar toekomst*, Hoofddorp/Ede: KPMG and SOAB.

Algemeen Verbond Bouwbedrijf (1996b) *Samen werken aan nieuwe arbeidsverhoudingen*, The Hague: AVBB.

Algemeen Verbond Bouwbedrijf (2000) *De bouw in cijfers, 1995–1999*, The Hague: AVBB.

Buurmeijer, F. (1993) *Parlementaire enquête, uitvoeringsorganen sociale verzekeringen*, The Hague: Tweede Kamer (Lower House of Parliament) 1992–3, 22 730, nos. 7–8.

Centraal Bureau voor de Statistiek (CBS) (various years) *National accounts*, Voorburg: CBS.

Economisch Instituut voor de Bouwnijverheid (EIB) (1994) *De bouwarbeidsmarkt, 1982–1994*, Amsterdam: EIB.

Economisch Instituut voor de Bouwnijverheid (EIB) (1997) *Uitbesteding door hoofdaannemers*, Amsterdam: EIB.

Economisch Instituut voor de Bouwnijverheid (EIB) (1998a) *De zelfstandigen zonder personeel*, Amsterdam: EIB.

Economisch Instituut voor de Bouwnijverheid (EIB) (1998b) *De personeelsvoorziening in de bouw*, Amsterdam: EIB.

Economisch Instituut voor de Bouwnijverheid (EIB) (1999a) *Bedrijfstakprofiel bouwnijverheid, 2000–2005*, Amsterdam: EIB.

Economisch Instituut voor de Bouwnijverheid (EIB) (1999b) *Het gespecialiseerde aannemingsbedrijf*, Amsterdam: EIB.

Economisch Instituut voor de Bouwnijverheid (EIB) (2000) *De bouwarbeidsmarkt in het najaar van 2000*, Amsterdam: EIB.

Eichhorst, W. (1998) *European Social Policy Between National and Supranational Regulation: Posted Workers in the Framework of Liberalised Service Provisions*, Working Paper No. 6, Cologne: Max Planck Institute.

Esping Andersen, G. (1996) *Welfare States in Transition. National Adaptation in Global Economies*, London: Sage.

European Court of Justice (1996) *Order of the Court of 25 March 1996, Vereniging van Samenwerkende Prijsregelende Organisaties in de Bouwnijverheid and others v Commission of the European Communities*, Case C-137/95, Brussels.

Federatie Nederlandse Vakbeweging (FNV) (various years) *Annual Review*, Woerden: FNV.

Hemerijck, A. C., van der Meer, M., and Visser, J. (2000) "From coordination to innovation: twenty years of social pacts in the Netherlands," in G. Fajertag and P. Pochet (eds), *Social Pacts in Europe: New Dynamics*, Brussels: European Trade Union Institute/Observatoire Social Européen, 257–78.

Hertogh, M. and Peet, J. (1999) *Werken aan uitvoering: de geschiedenis van de bedrijfsvereniging voor de Bouwnijverheid en het sociaal fonds bouwnijverheid*, Amsterdam: NEHA/SFB.

Interview with employers' representative, VG-Bouw, Zoetermeer, 20 January 1995.

Jansen, I., Jeurissen, R., and Lenoir, P. (2001) *Kiezen en delen, de ontvlechting van de SFB-groep*, Amsterdam: SFB.

Korevaar, K. (1990) *Funderingsherstel*, Amsterdam: Uitgeverij Mets.

Ministry of Social Affairs and Employment (1998) Internal database on extension of collective agreements, The Hague.

Scholingsfonds (various years) *Annual Report*, Amsterdam.

Sociaal Fonds Bouwnijverheid (SFB) (various years) *Annual Report*, Amsterdam: SFB.

Sociaal Fonds Bouwnijverheid (SFB) (1998) *Yearbook for employers*, Amsterdam: SFB.

Stichting Vakopleiding Bouwnijverheid (SVB)/Bouwradius (various years) *Annual Report*, Zoetermeer: SVB.

te Lintel Hekkert, K. (1992) *Verleden en perspectief, 25 jaar opleidings- en ontwikkelingsfonds*, Amsterdam: Stichting opleidings- en ontwikkelingsfonds.

Unger, B. and van Waarden, B. F. (1993) "Construction industries: a comparison between Austria, Germany, Great Britain, Italy, the Netherlands, Sweden and Switzerland," in M. Ledoux and P. Llerena (eds), *The future of industry in Europe*, Brussels: Commission of European Communities.

van der Meer, M. (1998) *Vaklieden en werkzekerheid, kansen en rechten van insiders en outsiders op de arbeidsmarkt*, Amsterdam: Thela Thesis.

van der Meer, M. (1999) "Labour market reform in the Dutch construction industry: lessons for Germany?" *Industrielle Beziehungen* 6 (3): 360–84.

van der Meer, M. (2001) "De modernisering van de arbeidsverhoudingen in de bouwnijverheid," *Sociaal Maandblad Arbeid* 56 (3): 175–94.

van Waarden, B. F. (1989) *Organisatiemacht en belangenverenigingen- de ondernemers organisaties in de bouwnijverheid als voorbeeld*, Amersfoort/Leuven: Acco.

3 Germany

The labor market in the German construction industry

Gerhard Bosch and Klaus Zühlke-Robinet

INTRODUCTION[1]

Both the product market and the labor market in the German construction industry are covered by a dense regulatory framework. Regulations of the product market are older than those of the labor market and reach back in part to the tradition of the trade guilds. Performance of construction work within so-called "full-craft" trades is confined to master craftspersons, who have undergone an officially recognized vocational training. Regulation of the construction labor market commenced as early as the end of the 1940s, at a time when the expansion of the welfare state was just beginning. The aim of this regulation was to mitigate skill shortages and the social problems experienced by construction workers as a result of the highly cyclical and seasonal nature of construction work. The high job insecurity of construction work has repeatedly led construction workers to seek jobs in other industries and has also caused a shortage in skilled workers. By introducing numerous supporting tools, such as a bad-weather allowance, an additional annuity, a supra-company vacation pay, and a levy collected for financing vocational training, the social partners attempted to sustain an industry-specific craft labor market. Most of the regulations solely affect construction workers in the so-called construction industry proper – i.e., in activities related to the erection of building frameworks and exterior shells. Neither in the completion work (i.e., interior work) nor in other areas of craftspersonship do we come across similar labor market regulations.

In contrast to other European countries with similar labor market regulations, collectively bargained arrangements of the trade unions and enterprise associations play an exceptionally important role in Germany. This became possible through the creation of an overall trade union for the construction sector after World War II and through the unification of the construction trade unions, which were formerly split according to trades, social status (blue-collar vs. white-collar), and political tendencies. Thus, collective bargaining agreements in the German construction industry now are being concluded for the entire sector and not just for individual craft divisions. This made it easier for the State to declare part of these agreements as generally binding – and for the social partners to speak with one voice in negotiating agreements on additional regulations with the State.

Until the 1980s, the regulations had a formative effect on the product and labor markets of the construction industry. Wages and social benefits had been widely removed from the list of competition parameters, which, owing to the great importance of quality standards and the high education and training standards applicable to masters, were instead focused on the quality of building work. This type of competition was beneficial to innovation and constituted an important reason for the high increases in productivity experienced by the industry throughout the 1960s and 1970s.

Since the end of the 1980s, this regulatory system has become fragile. Several developments fostering this course came together at that time. In the wake of freedom of services being introduced in the EU, and due to the use of posted construction workers from Central and Eastern Europe, workers not paid according to the German wage system were working on German construction sites. Simultaneously, a considerable wage gap opened between West and East Germany after German reunification. Consequently, the wage aspect again became an essential competition parameter, all the more because the construction industry began to decline, developing from a seller's to a buyer's market. German reunification had further indirect effects: Due to the reconstruction aid for East Germany, the State got into financial difficulties and, because of EU membership criteria set forth in the Maastricht Treaty of 1992,[2] it was required at the same time to reduce national debt. The State therefore sought to withdraw from the labor market policy for construction workers.

The social partners, the Chambers, and the State have reacted to these different challenges. Over the past years, the regulations of the product and labor markets have undergone considerable changes, and they have been developed further – however, without a new stable condition having been achieved so far.

In this chapter, we shall have a closer look at the construction labor market in Germany. We shall begin with an analysis of the product market and the labor market in the German construction industry. Then the regulations in both markets will be analyzed. The modification of the regulatory system of the labor market within the past years will be examined, and the final section will summarize findings.

ECONOMIC EVOLUTION

The role of the construction industry in the overall economy

In 2000, slightly more than every seventh German mark was spent on construction work. A construction volume on the scale of 531 billion German marks stands against a GDP of 3846 billion marks, which corresponds to a share of 13.8 per cent. Since much preliminary work in the field of construction is purchased from other branches of industry (such as building materials, prefabricated constructional elements, and building machinery), the share of gross value added in the construction industry was much lower, leveling out at 5.3 per cent. The construction volume is thus almost three times as large as the amount of gross value added in the construction industry.

The development of the construction industry in West Germany has taken a very different course from that in East Germany. In West Germany, the share of gross value added of the construction industry fell from 10 per cent in 1960 to 5.3 per cent in 2000 (see Table 3.1).

There are four main reasons explaining the declining importance of the construction industry for the overall economy in West Germany. The first reason is that the reconstruction of towns and cities, of infrastructure, and of production plants destroyed in the War had already been completed to a large extent in the 1970s. Second, manufacturing plants today require less enclosed space, which is due to changes in production concepts. Third, the growing service sector invests less in buildings than the manufacturing sector does. Fourth, financial bottlenecks experienced by public contract awarders have led to a decline in construction activities. The share of public demand for building work has decreased from over 30 per cent in 1970 to less than 15 per cent in 2000.

The situation in Eastern Germany is very different. The share of the construction industry in gross value added was 18 per cent in 1995 – three times larger than it was in West Germany. The former German Democratic Republic (GDR)

Table 3.1 Share of the construction industry in gross value added and share of the construction volume in the gross national product between 1960 and 2000, in per cent

Region/ Country	Year	Share of the construction industry in gross value added	Share of construction volume in the gross national product
West Germany	1960	9.8	20.0
	1990	5.9	15.0
	1991	5.6	13.4
	1995	5.3	14.2
	1998	no data available	12.6
	1999	no data available	12.4
Eastern Germany	1991	12.4	19.7
	1995	18.2	32.0
	1998	no data available	27.7
	1999	no data available	26.6
Germany	1991	6.4	14.0
	1998	5.9	14.5
	2000	5.3	13.8

Sources: IG BAU 2001; Bosch and Zühlke-Robinet 2000: 35; authors' calculations.

Notes
1960 and 1990 at prices of 1991; from 1991 at prices of 1995; since 1998, data on value added per industry have no longer been shown separately for West and Eastern Germany.

survived for decades on existing resources – i.e., on the substance of its infrastructure and of its companies, which often dated from the 1920s and 1930s – and therefore there was considerable demand for restoration, renovation, and refurbishment following German reunification. The construction trade was one of the driving forces behind the high economic growth rates in Eastern Germany in the period between 1991 and 1995. The year 1996 marks the economic turning point within the construction industry of Eastern Germany. With the gradual end of the reconstruction era and the phasing out of the generous tax allowances for both building and capital investors, the economic weight of the building sector started to decline. In future, the number of employees will continue to decline more strongly than construction contracts because East German companies will endeavor to catch up on productivity shortfalls and try to approach the West German level.

Thus, now the building sector is shrinking in both parts of Germany. As we know from economic theory, price competition increases in shrinking markets. Since capacities exceed demand by far, the German construction industry has undergone a change from a supplier's market to a buyer's market. Since 1996, the price of shell erection has been decreasing for the first time since the middle of the 1960s.

The construction industry in the economic cycle

Between 1951 and the end of 1997, construction volume increased by an annual average of 3 per cent, whereas the overall economy grew by an annual 4 per cent over the same period. Cyclical trends are smoothed out in such average figures, and therefore turbulent courses of economic business cycles are difficult to detect. However, a look at the course of these figures over a year's period reveals that the cyclical and seasonal peaks and valleys of construction volume are much more pronounced than those indicating overall economic growth (see Figure 3.1).

Changes in company structures

Over the past decades, nearly all of the larger German building firms have developed into general contractors and building service companies. They have increasingly taken over the complete handling of construction orders, which are partly executed by their own personnel and partly by subcontractors. Moreover, several building firms have extended their value-added chain by integrating further up- or downstream activities into their range of products and services offered. The spectrum here stretches from the supply of landed property, project development and financing, drafting and planning work via the entire construction process to building marketing and the most favorable organizational, technical, and commercial management of buildings.

The expansion of the scope of business as well as the internationalization of large building firms is leading to a supply split in the building trade (Rußig, Deutsch and Spillner 1996). The more numerous the functions taken over by

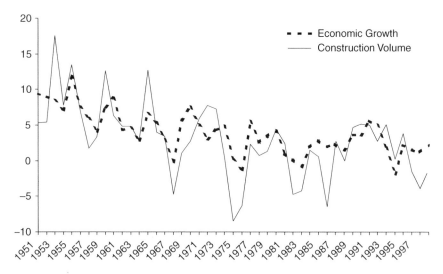

Figure 3.1 Construction volume and economic growth in West Germany between 1951 and 1997, changes as against the preceding year in per cent.

Source: Bosch and Zühlke-Robinet 2000: 41.

larger building firms and the more technical and organizational competence they concentrate on themselves, the more small and medium-sized companies are likely to assume the subordinate role of subcontractors. Next to the classic type of regionally oriented, independently operating small and medium-sized companies, focusing for instance on the construction of new residential buildings, we now see hierarchically shaped buyer–supplier structures developing between large-scale companies and small firms. Structures of this type have been known a long time in the processing trade. So far, only a few small building firms have attempted to stem this growing dependence on general contractors by joining forces in suppliers' associations designed to offer a wide range of pooled services and products.

Following this extension of the value-added chain, or workbench extension, the big building companies have noticeably changed their personnel structures. In the 1960s, they employed seven industrial workers per one white-collar worker, whereas the ratio now is down to one-to-one in large-scale companies. Big establishments today develop small, stable core workforces and otherwise tend to externalize employment fluctuations. Especially cycle- and weather-sensitive industrial activities such as shell building are now purchased from subcontractors. Since the end of the 1980s, this development has been on the increase, and it is being further accelerated by the involvement of foreign subcontractors whose labor force is prepared to work for lower wages.

The consequence is a growing segmentation of supplier structures. A smaller number of large-scale enterprises controlling long value-added chains confront a growing number of small and medium-sized companies. Therefore, the average company size in this sector, which has at all times been dominated by small and medium-sized firms, has further decreased over the past decades. The average company size in the West German building industry declined from 21 staff members in 1976 to 13 in 2000. Forty years ago, there were still approximately 130 companies each having more than 500 employees in West Germany; the figure has since then diminished to about 30. Following the dissolution of the huge former construction combines in Eastern Germany, company size structures there, are approaching the West German level. The increased importance of small and startup businesses is also reflected in the absolute number of building firms. Despite the lower economic weight of this industry in West Germany and despite the decline in employment, the number of building companies has remained largely constant over the past decades, leveling out at approximately 60,000 companies. At the end of June 2000, there were 59,262 companies in West Germany; in 1970, their number had amounted to 63,415; and the 1955 figure was 62,836 (Hauptverband der Deutschen Bauindustrie 2001; IG BAU 2001).

By contrast, a considerable increase in the number of companies can be observed in Eastern Germany since 1991. In 1991, there were only 7000 firms in Eastern Germany, whereas in 2000 the figure already exceeded 21,850 (IG BAU 2001). It is less a growth in the demand for building work that is reflected by these figures than the dissolution of large construction combines and industrial restructuring in the sector.

The growing involvement of subcontractors also becomes visible in the cost structures of building companies. In West Germany, expenses defrayed for hired labor more than doubled in the period between 1970 and 1999. In the year 1970, the share of subcontracted work in turnover was approximately 12 per cent, whereas in 1999 it amounted to almost 30 per cent. In 1999, 8 per cent of sub-contracted work was accomplished by foreign subcontractors (IG BAU 2001: 30).

EMPLOYMENT AND UNEMPLOYMENT

By 2000, the number of people employed in the West German construction industry had declined by 1,000,000, or 58 per cent, from its 1964 peak (see Table 3.2). While in 1964 the share of employees in the construction industry proper was 8 per cent, this portion dropped to 3.1 per cent in 2000. Eastern Germany is undergoing a similar development after German reunification; however, in Eastern Germany this change is happening in a time-delayed manner. Until 1995, the number of employees in the construction industry proper increased by over a third, and it has been decreasing again ever since. However, the share in over-all employment of those employed in the construction industry continues to be twice as large as in West Germany (Bosch and Zühlke-Robinet 2000: 70).

The employment structure has also undergone considerable changes. Workers without vocational certifications were the category worst affected by this

Table 3.2 Employment structure in the construction industry

Year	No. of employees (1000's)	Employee Category (as percentage of total)*				
		Skilled tradesmen including foremen	Commercial and technical personnel	Ancillary and unskilled workers	Working owners, co-owners and active family members	Apprentices**
West Germany						
1950	912.8	48.9	5.0	36.6	8.6	10.2
1964	1720.0					
1990	1033.6	61.2	15.3	18.1	5.0	4.0
1995	982.6	58.9	17.7	18.6	4.4	5.8
2000	720.8	56.2	20.0	17.5	6.2	5.1
Eastern Germany						
1991	327.2	68.0	18.1	12.0	1.7	7.5
1995	450.8	62.6	15.1	20.1	2.1	8.6
2000	288.7	59.4	17.4	17.7	5.5	6.7
Germany						
1991	1396.4	62.5	16.3	16.9	3.6	4.5
2000	1009.5	57.1	19.3	17.6	6.0	5.6

Sources: Bosch and Zühlke-Robinet 2000: 71, 76; Statistisches Bundesamt 2001: 35–7; authors' calculations.

Notes
* Excludes trainees.
** Craft apprentices only.
New method of recording data for the construction; industry introduced in 1995; only restricted comparisons with previous years possible.

reduction in employment. Whereas in 1970 there were two skilled workers for every unskilled worker, by 2000 this ratio had increased to a 3:1 proportion. Studies of work organization on building sites have shown that unskilled tasks are being added to the range of tasks undertaken by skilled workers. Construction firms are increasingly relying on teams of skilled workers able to work in a largely independent manner (Syben 1999). In marked contrast to other countries (e.g., the United Kingdom), the craft structure of the German construction labor market has thus been strengthened in recent decades. The stabilization of vocational training in the industry after a temporary crisis has contributed significantly to this development.

The share of white-collar workers has increased significantly. In 1950, only one in 20 employees was a white-collar worker; by 2000, one employee in five was in that category. This development is a consequence of increasing operational complexity. The greater the effort devoted to customer canvassing, the stronger the need for cooperation among the various companies engaged in construction projects and, significantly, the greater the advances in construction and machine technology that make work planning and scheduling tasks increasingly demanding. In contrast to many other countries, especially the United Kingdom, the share of self-employed workers (including helping family members) is at 6 per cent and thus very low. The German Handicrafts Code, which allows only master craftspersons to establish construction firms, exerts a brake on the self-employment rate. Thus, the proliferation of independent contractors that has been so widespread in the United Kingdom is in Germany necessarily limited by the qualifications associated with master craftspersons. Nonetheless, owing to the creation of many very small companies and the dissolution of the former Eastern German combines, the self-employment figure more than doubled between 1991 and 2000.

The construction labor market is characterized by high external turnover and unstable employment. Figure 3.2 shows that the turnover rate in the construction industry is considerably higher than that in the economy as a whole. This high turnover and the instability of employment relationships in the construction industry are primarily a consequence of the strongly cyclical and seasonal nature of the industry. To this is added a heavy dependence on the weather. Bad weather in winter often means that work on existing orders has to come to a halt or cannot even be started. Moreover, seasonal fluctuations are further exacerbated by the time constraints imposed during order placing: fearing an onset of bad weather, customers request fulfillment of their orders during the

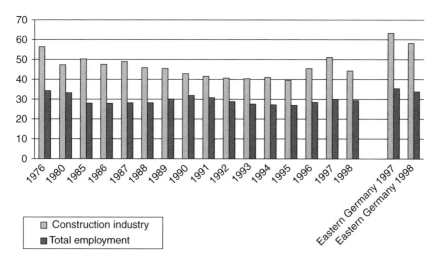

Figure 3.2 Average annual turnover among workers in insurable employment in West and Eastern Germany, as percentage. Defined as number of new hires liable for contribution to social insurance (within a given year)/2x employment (on a fixed date or on an annual average).

Source: Bosch and Zühlke-Robinet 2000: 87.

time of good weather, between April and October. Orders decline in winter, and employment has to be reduced accordingly. Finally, the high share of small firms in the construction industry is also important in this respect, as small firms have virtually no resources to retain workers in slack periods.

The other side of the same coin is that a rapid cutback in employment levels can be followed by a high short-term demand for labor. If employment picks up in the spring or if a spell of bad weather comes to an end, firms are eager to recruit again. The construction industry thus needs some labor in reserve, a pool of unemployed persons who will not drain away into other industries during temporary unemployment. In many cases, firms hire at least some of the workers they had previously dismissed. There are often tacit agreements to the effect that employed persons will resume work for the same company after a period of unemployment in the winter. Therefore, it is common practice in the construction industry for many redundant workers to receive recalls from their previous employers (Mavromaras and Rudolph 1995).

One important characteristic of unemployment in the construction industry is its marked seasonality. Unemployment increases sharply in winter and declines rapidly in the spring. Until the introduction of the statutory bad-weather regulations and their incorporation into collective agreements at the end of 1959, the problem of seasonal unemployment used to be part of the construction worker's lot. Periodically recurring winter unemployment rose rapidly winter after winter, reaching a new peak of over 800,000 in the winter of 1955–56. Figure 3.3 clearly shows the characteristic seasonality of unemployment in the construction

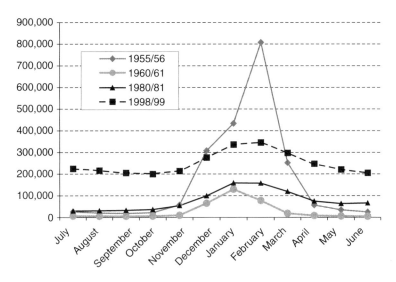

Figure 3.3 Seasonal unemployment in construction jobs, selected years since 1955 and 1998/99 for total Germany.

Source: Bosch and Zühlke-Robinet 2000: 178.

industry; due to the introduction of the bad-weather allowance in 1959, the labor reserve was in part internalized, and the level of seasonal unemployment is considerably lower than it was in 1958 and before.

Since the beginning of the crisis in the construction industry, unemployment has substantially increased, with marked regional differences between West and Eastern Germany. A West German peak in unemployment had already been reached in 1997, at almost 140,000, declining by the year 2000 to 100,415, whereas unemployment in Eastern Germany has increased over the years, reaching a level of over 117,000 in 2000, up from 35,333 in 1994 (IG BAU 2001).

REGULATIONS IN THE PRODUCT MARKET

Protection and role of the master craftsperson title

The most important provisions relating to the protection of the master craftsperson title can be found in the Handicrafts Code (*Handwerksordnung*). The Code went into effect in 1953, but its customary and legal roots were largely in place by the nineteenth century, when guilds monitored the quality of handicraft services and permitted the establishment of trades. With the adoption of the Industrial Code (*Gewerbeordnung*) in 1897, the work of the guilds was placed on a legal footing. At that time, the law also provided for an establishment of regional Handicraft Chambers (*Handwerkskammer*). The Handicrafts Code specifies with binding force who is allowed to set up and run a handicraft business. Here, a distinction is made between "full-craft" and "quasi-craft" trades. In contrast to the quasi-craft trades, full crafts may be pursued only by someone with a master craftsperson's certificate. "Annex A" of the Handicrafts Code lists 94 different trades as "full crafts," for which there is specific training. In the field of construction, a delimitation comprising 18 trades has been set up (see Table 3.3).

Table 3.3 Index of trades belonging to the group "Construction and Outbuilding Trades," which may be exercised either as a full craft (Annex A) or as a quasi-craft trade (Annex B)

Annex A

Bricklayer and Concrete Worker; Stove and Air Heater Manufacturer; Carpenter (including for prefabricated houses, tilted roofs and dry lining); Roofer (including flat roof sealing and roof truss); Road Construction Worker (underground works and canals); Thermal and Acoustic Insulation Worker (including dry lining); Tile, Slab and Mosaic Setter; Manufacturer of Precast Concrete Blocks and of Terrazzo; Screed Worker; Well Sinker; Stone Mason and Stone Sculptor; Plasterer (including dry lining), Painter; Scaffolder; Chimney Sweeper.

Annex B

Bar Bender; Construction Drying Trade; Floorer; Asphalt Worker; Jointer in Building Construction; Timber and Floor Preservation; Pile-Driving Trade; Concrete Borer and Cutter; Theatre Equipment and Scenery Painter.

Registration with the municipal authorities and enrollment in the Register of Craftspersons (*Handwerksrolle*), which is maintained by the local Handicrafts Chambers, is imperative for anyone wishing to run a full-craft business. Enrollment in the Register is restricted to self-employed individual entrepreneurs who have passed their master examination (or an officially recognized engineer examination) in the craft exercised by them or in a craft declared as related. The examination may be taken after a 3-year period at a journey status and requires evidence of practical vocational skills, theoretical knowledge in the respective discipline, business management and legal knowledge, and knowledge in the field of employment and work pedagogy. The master craftsperson title allows the holder to train apprentices in the registered trade.

When construction companies in the industrial sector perform full-craft tasks, they are obliged to employ appropriately trained masters. In 1999, approximately 77 per cent of the building companies, accounting for 73 per cent of construction workers, were registered as pursuing a handicraft trade. The question of whether a company may be a member of the Chamber of Handicrafts (*Handwerkskammer*) or of the Chamber of Industry and Commerce (*Industrie- und Handelskammer*), is not a matter of size, turnover, or the focus of a company's activity, but rather depends on who founds or runs the enterprise.

In light of the vocational mobility being implemented throughout the EU, the German labor market, formerly closed to foreign handicraft services by the requirements of its training system, had to be opened. To preserve vocational training in Germany and the central role of the master craftsperson title, the market was opened only to qualified workers. Only EU citizens providing official certificates proving that they have pursued the respective handicraft, either on a self-employed basis or in a leading position, in their home country for a duration of at least 6 years – and for the last time no longer than 10 years ago – may now be enrolled in the Register of Craftspersons and become self-employed in that trade in Germany.

The motivation behind the protection of the master craftsperson title and the delimitation of the various trades is quality assurance. A person who performs a trade without being able to present the required master craftsperson's certificate infringes the Handicrafts Code and thus the Law against Illicit Work and may be liable for a fine of up to €100,000. In Germany, this quality assurance is sought, as in the case of freelance professions (physicians, lawyers, etc.), by means of a generally recognized certification. Since technical progress will continue to change established definitions, this objective can be achieved only if current trade demarcations are revisited at regular intervals and revised if necessary. This was last done in Germany in 1998, when the Handicrafts Code, among others, was amended by uniting certain trades, with new demarcations being established at the same time, leading to 94 full-craft and 57 quasi-craft trades. A big obstacle to such revisions is the traditional way of thinking in vocational categories, which is typical of craft guilds – i.e., of the local and regional associations of individual trades.

Transparency, quality assurance, and invitation to tender

The construction market is extremely nontransparent to customers because its products are hardly standardized. This holds much potential for conflict. Comparability of offers is thus rendered difficult, and disputes about quality and price often arise when it comes to the acceptance or surveying of accomplished construction work. To defuse this potential source of conflict and increase market transparency of public contracts, the German Building Contract Code (*Verdingungsordnung für Bauleistungen*, or VOB) was established, based on which, virtually all construction work is handled. Its roots stretch back into the 1920s. Following World War II, the Code was continually revised by the German Association for Construction Contract Procedures (*Deutscher Verdingungsausschuß für Bauleistungen*, or DVA). Currently, this institution has about 50 member-authorities and organizations.

Part A of the VOB contains principles applicable to invitations to tender, specifications, the tendering or bidding process, and the awarding procedure. It stipulates that all work to be performed by the building companies shall be described in detail so bidders can submit realistic quotations. It also lays down that the different services to be furnished shall be put out to tender separately, according to the various trades involved, to enable small and medium-sized companies to participate in the bidding process. The tendering period for building firms is fixed. On principle, the contract shall be awarded to the bidder submitting the most favorable quotation in light of company integrity and reliability of offer.

According to this set of rules, the public sector is principally obliged to issue invitations to tender for all construction orders, except extremely small orders. If the order amount exceeds €5 million, the stipulations of the EU Construction Coordination Directive of 1989 must be observed. The Directive aims at making the public tendering and contracting practice more transparent, by providing for longer tendering periods and uniform tendering and contracting procedures for all companies within the EU. Moreover, the VOB rules stipulate the time following the conclusion of building contracts. Their main aim is to ensure smooth and proper accomplishment of construction work. They are also meant to help solve conflicts arising, for example, between clients and contractors, if need be. The final part of VOB covers the technical terms and conditions applicable to building contracts. Building contracts also have to follow the stipulations of the EU Building Product Directive of 1988, which is intended to ensure harmonization of standards within the EU. This Directive defines the major constructional requirements – stability, fire protection, and environmental protection – that must be taken account of in national directives.

The VOB has the status of neither law nor ordinance, but it has definitely become common practice for contract awarders and contractors to shape their contractual relations based on the VOB. Public authorities have undertaken to subject all of their building contracts to the VOB. The VOB has thus developed into the central tool for quality enhancement and for increasing market transparency. It further constitutes a means of reference in the settlement of disputes between contract awarders and contractors. Since the technical standards are revised and amended at regular intervals, they are also a dynamic factor in the field of product design.

Finally, the EU Building Site Directive of 1992 must be mentioned. In 1998, the Directive was transferred into German law. The Directive aims at enhancing safety and health protection of construction workers by means of improved planning and coordination.

THE REGULATION OF THE CONSTRUCTION LABOR MARKET

The parties to collective bargaining

The organization of employee interest representation is relatively simple, since there is only one trade union active in the construction industry, the Trade Union for Building, Forestry, Agriculture, and the Environment (*Industriegewerkschaft Bauen–Umwelt–Agrar*, or IG BAU). It represents employees in the construction industry proper and virtually all trades in craft establishments; with some 615,000 members as of December 1998, it is the fourth largest union in the German Trade Union Federation (*Deutscher Gewerkschaftsbund*). Approximately 40 per cent of construction workers are organized in the IG BAU. Since collective agreements are negotiated centrally, collective bargaining policy is the responsibility of IG BAU's national committee.

Interest representation on the employers' side is less monolithic. There are two employers' umbrella organizations. The industrial sector is represented by the Central Association of the German Construction Industry (*Hauptverband der Deutschen Bauindustrie*, or HDB), while the handicraft sector is covered by the Central Association of German Building Trades (*Zentralverband des Deutschen Baugewerbes*, or ZDB). Each of the two organizations is a "federation of federations" whose members are formally independent federations covering the individual federal states (*Länder*). The HDB has a total membership of 16 *Länder* and inter-*Länder* employers' federations, while the ZDB has 41 *Länder* and regional federations, which in turn are based on regional craft corporations and regional associations.

In contrast to other industries in Germany, the authority to negotiate in the construction industry is concentrated at the central level on both the employers' and the employees' side. This creates the prerequisite for negotiations on industry-wide regulations. Each of the central collective bargaining bodies in the umbrella organizations must be in a position to bring together the heterogeneous interests of the various organizational units below that central level in order to construct a consensus (Weitbrecht 1969). IG BAU's internal structures are better suited to this task than those of the two employers' umbrella organizations, for the regional and local subdivisions of IG BAU do not have the option of leaving the federation and entering into collective bargaining on their own authority.

Collective agreements

One fundamental characteristic of labor relations in the construction industry is the centrally negotiated and concluded collective agreements. These provide the basis for industry-wide regulations and are also considered necessary by the representative associations on both sides as a means of ensuring that common standards apply both to construction companies operating outside their own regions and to workers from several firms in different regions brought together to form work teams.

Since not all firms are members of the employers' federations, the only way of preventing regulations from being undermined by "unfair competition" from outsiders is to declare collective, generally binding agreements.[3] Some of the most important collective agreements to have been declared generally binding are the federal outline collective agreement for the construction industry, which contains provisions on job classifications and allowances for work outside the local region, the collective agreements on social funds, and the collective agreement on minimum wage rates. In contrast, none of the agreements on wages and salaries has been declared generally binding. Thus, regional differences of wage levels are permitted. In daily practice, these differences were of minor importance until the beginning of the 1990s, because the trade unions negotiated similar wages in all regions. Since German reunification, however, there have been substantial regional differences on the German construction labor market (for example, standard wages in Eastern Germany are 15 per cent below those paid in West Germany), so wages have once again become a competition parameter.

As of January 1, 1982, legislation prohibited temporary agency employment of blue-collar workers in the construction industry. The opinion was that temporary labor agencies jeopardized the "good order" in this part of the labor market as well as the social security of those working in it. This prohibition aimed to safeguard the regulatory function of the generally binding collective bargaining agreements in the construction industry and to eliminate competition from hired personnel.

The social funds

At the end of 1948, the first social fund for the payment of wages during vacations in the construction industry was established by collective agreement. Firms paid a certain percentage of their wages bill into the fund, which was then used to pay wages when vacations were taken. This arrangement enabled employees to receive vacation pay even if they changed jobs several times in the course of a year. In calculating individual vacation entitlements, all periods of employment in construction firms were aggregated. The year 1955 saw the introduction of a wage compensation fund that financed wages between 24 and 31 December through a levy to protect workers from being laid off before Christmas. This vacation and wage compensation fund (*Urlaubs- und Lohnausgleichskasse*) was joined in 1959 by the supplementary pension fund (*Zusatzversorgungskasse*), an industry-wide occupational pension plan for the construction industry, which made it more attractive for employees to remain in the industry.

In 1974, the funds became responsible for financing the vocational training of construction workers. During the 1960s, the number of apprentices had declined sharply; by 1970, the share of apprentices in total employment in the industry was only 1.8 per cent, compared with 10.2 per cent in 1950. The old, craft-oriented training system was clearly no longer able to provide a good supply of young workers to replace an aging workforce. To reverse this decline, the social partners decided to adopt completely new methods through a reform of the training system. Following this reform, training took place in two stages, with the second building on the skills and knowledge acquired in the first; a vocational certification could be acquired at the end of both stages. In the first year in particular, training went beyond the confines of individual trades. In order to ensure this, apprentices spent alternating blocks of time in external training centers serving a number of firms and in vocational schools. The reform also changed the way in which such training is organized. The time spent by apprentices in firms fell from 80 per cent to 46 per cent of their working time, the time spent in vocational schools rose from 20 per cent to 26 per cent, and the time spent in the joint training centers is around 20 per cent (see Figure 3.4).

In order that the burden of the higher costs[4] of the improved vocational training system should not weigh too heavily on firms, a financing system based on levies and administered by the vacation and wage compensation fund was introduced.

1. Bricklayer
2. Concreate and armored concrete builder
3. Furnace and chimney builder
4. Carpenter
5. Concrete stone and terrazzo
6. Stucco maker
7. Tiler
8. Floor finisher
9. Insulation builder
10. Dry construction assembler
11. Road builder
12. Pipeline builder
13. Canal builder
14. Well builder

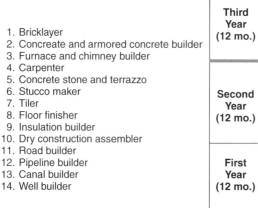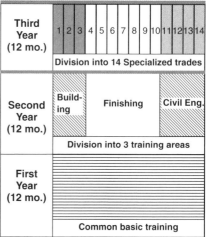

Figure 3.4 Vocational training system: Integrated Training Scheme for the 14 construction trades.

Source: Streeck *et al.* 1987: 63.

All firms in the construction industry pay a monthly contribution to a fund that is used to reimburse a considerable proportion of the costs incurred by those firms that offer training places for apprentices. In 1975, this was 0.7 per cent of gross wages; in 2000, 2.8 per cent. Training firms receive a refund of their vocational training costs.

The success of the reform and of the system of financing is irrefutable. The training ratio – i.e., the share of apprentices in overall employment within the construction industry proper – rose from 1.8 per cent in 1970 to over 6 per cent at the end of the 1990s. The reform was accompanied by a significant increase in training allowances for apprenticeships in order to make the building trades more attractive to young people.

The two social funds are administered jointly by the three central federations empowered to engage in collective bargaining – i.e., the trade union and the two employers' federations. In 1998, the funds employed around 1300 people and had a balance sheet total of €5.8 billion. This makes them an important economic factor that helps to sustain the close links between the social partners. The funds are considered "a kind of training organization where the way the collective bargaining partners should behave towards each other, namely in a humane and prudent way, is being practiced" (Fürstenberg 1988: 43).

Most of the benefits can be obtained by blue-collar workers only. White-collar workers receive only their additional annuity from the social funds. The contributions of industrial workers depend on the companies' gross-wage sums, whereas a per-capita premium must be paid for white-collar workers. The level of contributions and the procedures for distributing them are laid down in collective agreements that have all been declared generally binding by the Federal Minister of Labor. At the beginning of the 1960s, company contributions to the social funds amounted to well over 11 per cent of gross wages. In the wake of benefit improvements, they increased until 1998 to 19.9 per cent in the former West German *Länder* and to 17.95 per cent in the former East German *Länder*. An amount of €30 per month is paid for every white-collar worker. Contributions made in the former East German *Länder* are lower because the benefits offered in West Germany have not been adopted in their entirety: the wage adjustment period comprises fewer days than in West Germany, and there are no collectively agreed supplementary benefits at all. Since vacation entitlement in the construction industry meanwhile amounts to 30 working days per year, the major share of the contributions is used to finance vacation pay – for example, it was 14.45 per cent in 1998.

Labor market policy

At the same time that the sphere of competence of the social funds was being extended in the 1960s, a number of labor market policy instruments to promote winter employment and safeguard jobs during the winter were introduced. The bad-weather allowance (*Schlechtwettergeld*) was a benefit amounting to 68 per cent of net pay paid to workers in compensation for loss of work and earnings due to adverse weather conditions during the statutory bad-weather period, 1 November to 31 March. The winter allowance (*Wintergeld*) was a bonus paid in addition to the standard hourly rate in recognition of the difficult conditions with which construction workers frequently have to contend in winter. The promotion of winter construction program (*Produktive Winterbauförderung*) was aimed at firms and was intended to recompense those firms that continued working during the winter for the extra costs they incurred thereby and/or make winter working more attractive to firms by offering them grants and loans. The bad-weather allowance was financed out of the unemployment insurance fund. The winter allowance and the promotion of winter construction program, on the other hand, were financed by a special winter construction levy on building firms. In 1975, this amounted to approximately 3 per cent of gross wages; in 1997, 1.7 per cent; since 2000, 1 per cent.

The bad-weather allowance in particular exerted a strong influence on the employment behavior of firms. It became financially attractive for them to retain workers during the winter, and the usual layoffs at the beginning of the bad-weather period declined considerably (see Figure 3.3).

CHANGES AND CHALLENGES

Until recently, the German construction labor market was a purely domestic market. All construction workers on German soil were covered by the collective agreements that had been declared generally binding. Furthermore, there was only a low regional wage differential, due to comparable levels of economic development. The main factors of change are the transnationalization of the German construction labor market and the gradual erosion of the regulatory system from within which followed German reunification. A further factor was the retreat by the federal government from the special labor market policy for the construction industry. Against a background of high unemployment, all these changes developed their potential to blow apart the regulations of the construction labor market.

The transnationalization of the German construction labor market

There has been a long tradition in the post-war period of employing foreigners in the German construction industry. At times, such as in 1976, they accounted for up to 15 per cent of all employees. Until the early 1990s, foreign workers were employed by German firms at the same rates of pay and with the same social and legal standards that applied to German workers. The freedom of entrepreneurs to provide services in another member state of the EU meant that workers from all EU member states could be posted from their companies to work on service contracts in Germany. The territorial principle that once governed the application of collective agreements no longer applies. Since posted workers in Germany

FORM	REGULATION OF WORKING CONDITIONS
Individual Migration	Territorial Principle
Posted Workers	Principle of Origin

Figure 3.5 Form and regulation of employment of foreign workers.
Source: Bosch and Zühlke-Robinet 2000: 215.

remain employees of the foreign company, they are employed on the terms and conditions prevailing in their home country (see Figure 3.5).

Since the beginning of the 1990s, the federal government has in addition concluded bilateral agreements with 13 Central and Eastern European countries on the use of posted contract workers. German firms may conclude contracts for building services with firms from these countries. It is on this basis that foreign firms may operate in Germany with their own workers. Each bilateral agreement lays down a different quota for the number of workers who may be posted to Germany. The agreements also stipulate that hourly rates of pay for these workers should take the collectively agreed German rate as a reference point, but that vacation entitlement, vacation pay, and sick pay are to be regulated in accordance with the laws of the country of origin.

Many posted workers from Eastern Europe and from EU member states have been working in Germany since 1987 (see Figure 3.6). Such employment reached its peak in 1996, when there were 188,000 posted workers in the German construction industry. Since at the same time unemployment among construction workers was beginning to rise, the posting of foreign workers received a good deal of criticism. As a result, quotas were reduced considerably, tougher controls were introduced to ensure, for example, that the rules on pay were being observed, and in employment office districts with high unemployment even stricter rules on the use of posted workers were introduced. Finally, in March 1996, the Posted Workers Act (*Arbeitnehmer-Entsendegesetz*) came into force.

The wages paid to posted workers are considerably lower than those of German workers. In some cases, the workers are being paid less than €5 per hour when the going rate in the area is over €10. There is an even greater difference in total hourly labor costs (including overhead), since posted workers do not contribute to the social funds. Employers reckon on hourly labor costs of €30 for German workers and between €18 and €20 for contract workers. In the case of the numerous illegal workers who have come into the German labor market in recent years, mainly from Central and Eastern Europe, the differences in costs are even more marked. Considerable numbers of German construction workers have been made redundant and replaced by contract or illegal workers.

The outsourcing of production has been considerably accelerated by the increased use of contract labor. Firms that rely solely on German workers are no

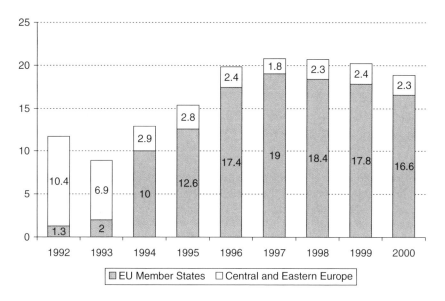

Figure 3.6 Share of posted workers from EU member states and from Central and Eastern Europe as percentage of all blue-collar workers in the German construction industry.

Source: Hauptverband der Deutschen Bauindustrie 2001: 25.

longer competitive and are forced to subcontract parts of their orders. Because of the traditionally high division of labor between firms in the construction industry, foreign subcontractors can be integrated quite easily into the construction process. However, since most posted workers do not have a vocational certification in one of the building trades, work organization is undergoing a process of "re-Taylorization," driven by the ready availability of cheap labor. The skilled German workers on building sites are increasingly required to take on a coordinating and supervisory role. Because of the serious problems with quality and the high costs of remedying poor work, German firms are trying, as part of their quality assurance programs, to develop stable relationships with reliable foreign firms. The advantages of using cheaper foreign labor seem to be greatest when a firm can dispense with specialist skills and high quality work – for example, in industrialized concrete construction and in so-called speculative building, where the simple, unfinished shells of buildings are purchased by companies using price as their main criterion (Syben 1999).

The fragility of the regulatory system in Eastern Germany

Whereas in West Germany the regulatory system in the construction industry has developed over decades, it was introduced in Eastern Germany overnight, upon reunification. East German enterprises were not prepared for competition, had a great deal of catching up to do in the technical and organizational aspects of construction, and were completely inexperienced in drawing up tenders. Their productivity levels were far below those of West German firms, while at the same time wages and costs of materials were quickly brought up to West German levels. Under these circumstances, paying less than the collectively agreed rate was, and still is, the only way most firms could and can survive. The high-level of unemployment in Eastern Germany meant there was an adequate supply of skilled labor prepared to work on such terms. In 1997, collectively agreed rates were being paid in only 45 per cent of construction firms located in Eastern Germany (Schäfer and Wahse 1998: 55).

Furthermore, firms have left the employers' federations, not only individually but also collectively. In the middle of 1997, the Berlin and Brandenburg Building Employers' Association resigned from the central federation because of disputes about collective bargaining policy. The main bone of contention was the rapid adjustment of wages in the East to West German levels. The tensions within the employers' camp, particularly between East and West, have increased considerably. For some time, pay bargaining was conducted on a joint basis. Since 1996, however, following pressure from the East German employers' federations, bargaining for Eastern Germany has been conducted separately.

New attempts to stabilize the construction labor market

Over the past years, the parties to collective agreements have jointly developed strategies with a view to making the regulatory framework of the lately crisis-prone construction industry weather- and crisis-proof again. The driving force behind this endeavor was undoubtedly the construction trade union, which wanted to prevent wage dumping. The union was supported by major sections of the employers' associations, who wished to achieve identical conditions of competition for all companies. Due to the close interweaving of collective agreements and political decisions, solutions from within the trades alone would not suffice. One of the prerequisites for success was that various politicians came on board again to agree on a joint course. In doing so, the social partners had to prove and confirm their will for reforms and not least of their readiness for sacrifice.

The most important changes in the regulation of the construction industry can be summarized as follows:

1 *Minimum wage rate for seconded workers*: The Posted Workers Act of 1996 allowed collectively bargained minimum wages for posted construction workers to be declared generally binding. The legislation also requires employers to register the posted workers they hire with the local employment office. The social partners agreed to a minimum wage of €9.30 in the West and €8.50 in the East. The representatives of the Confederation of German Employers' Associations (*Bundesvereinigung Deutscher*

Arbeitgeberverbände, or BDA) on the collective bargaining committees were unwilling to agree to minimum rates at such levels. The construction industry employers' federations threatened to resign from the BDA if agreement was not reached. After a lengthy debate, the BDA agreed to a minimum rate of €8.50, which was reduced in 1997 to €8 (€7.60 in Eastern Germany). In September 1999, minimum wages rose to €9.25 in the West and €8.10 in the East. This latest noticeable increase became possible only after the new Red–Green federal government had changed the process of declaring collective agreements generally binding. Now the Federal Ministry of Labor may determine by right of ordinance that a collective agreement on minimum wages shall be declared binding. The BDA thus lost its influence on shaping the minimum wage rate per hour.

2   *Statutory bad-weather allowance and flexible annual working hours*: The bad-weather allowance was abolished at the end of 1995, because of the poor financial situation of the federal government, particularly the Federal Labor Office. It was difficult for the social partners to find new arrangements to cover the loss of working hours in the winter; after the withdrawal of the State, both the employees and the companies had to bear more of the financial burden than before. Currently, the cost is being shared by the employees, the companies, and the State. The arrangement provides for an accumulation of overtime hours between spring and autumn that may be used to compensate for the first 30 hours lost due to bad weather. A loss of working hours from the 31st up to the 100th hour is being financed by means of a statutory winter construction levy payable by all companies. From the 101st hour onward, the loss is compensated for by the statutory unemployment insurance (with equal financing by all companies and employees). Consequently, industrial employees receive a stable monthly income over the year.

3   *Cost relief for businesses*: Between 1996 and 1999 – i.e., after the post-reunification boom in the construction industry – the partners to collective bargaining substantially reduced the companies' cost of labor. This was mainly to enhance their competitive edge in vying with foreign bidders. The collectively agreed annual wage increases fell far below former years' levels. In addition, these wage increases were counter-financed by cuts in fringe costs (i.e., vacation pay, sick pay, and the training allowance). Wage rates in the East were frozen at 94 per cent of the West level, and plans to equalize wage rates in the East and West were abandoned. Of greater significance is a *de facto* cut in wages of up to 10 per cent. This was dressed up as a "job protection measure," but it can be and is being implemented in all firms, since East German construction firms do not have to disclose their economic situation to works councils or the trade union. In most cases, however, this simply means that the widespread practice of undercutting agreed rates is being legalized. At the same time, these concessions stabilized the internal cohesion within the employers' associations and counteracted "wild flexibilization," which had already reached an advanced stage in the companies (Artus, Schmidt and Sterkel 1998: 126).

4 *Reform of the social funds and new benefits*: Since border-crossing construction work will continue to increase, the German social funds have concluded agreements with the social funds of several neighboring countries – the Netherlands, France, and Denmark. According to these agreements, construction workers posted from these countries are exempt from paying contributions to the German social funds, provided they can prove that they are paying contributions to the respective funds in their home country. This is a first step towards a border-crossing regulatory system. Moreover, the awkward administration regarding the collection of contributions and the equally tedious practice of contribution refunding are simplified. In the future, the funds will set off contributions received and refunded and thus avoid unnecessary payment transactions. Finally, the funds are taking over new tasks in the introduction of a new privately financed additional annuity. By means of these further functions and administrative simplifications, the social partners wish to strengthen the companies' acceptance of the funds.

5 *Measures to strengthen compliance with the law and with collective agreements*: The extension of subcontracting chains has made it increasingly difficult to safeguard compliance with the laws in force – i.e., minimum wages for posted workers and payment of taxes. Companies complying with the applicable laws and collective agreements have suffered a competitive disadvantage. To put a halt to the increasing illegal behavior within the construction industry, particularly tax evasion by subcontractors, the German Parliament passed a law in May 2001 stipulating that main contractors must withhold 15 per cent of the order value from their subcontractors and pay those amounts directly to the tax office. Several *Länder* – Berlin, Bavaria, Saarland, and Saxony-Anhalt – have passed so-called "contract allocation laws" (i.e., prevailing wage laws), restricting public contracts to companies complying with the applicable collective agreements. Evidence of such compliance must be provided by submitting a wage and salary roll attested by a notary or a competent Chamber proving that the negotiated wages are being paid or by a letter of the Works Council to that effect. On application of several *Länder*, the government drafted such an allocation law, also in May 2001, stipulating federal validity; however, it has not yet been passed.

CONCLUSION

The German construction labor market is covered by a dense regulatory framework, which aims at ensuring employee qualifications and quality of performance. The master craftsperson's monopoly ties the management of building firms within the handicraft trades to proof of qualification and thus constitutes an important barrier to an increase in the number of low-qualified self-employed people, existing, for example, in the United Kingdom. With the help of the VOB, or German Building Contract Code, trade-wide quality standards have been established. Although these are binding for public contracts only, they have a strong effect on private business practice as well. The regulations of the construction labor market, especially the generally binding collective bargaining agreements, the social funds, and the national labor market policy for the construction industry, pursue the aim of maintaining a pool of skilled labor by ensuring vocational training and increasing employees' ties to the industry. Towards these ends, the social problems caused by high inter-firm mobility, fluctuating demand, and adverse weather conditions have to be mitigated. The resolution of social problems in the construction industry is, therefore, a precondition for its economic functionality. If the job insecurity caused by fluctuating demand and adverse weather conditions cannot be partially resolved, then skilled workers will leave the industry and skill shortages in the construction industry will develop. A recent empirical study shows that the regulatory system contributes to tying qualified construction workers to their trade, contrary to the widespread assumption that skilled labor leaves the construction industry each year in higher percentages than in other industries (Erlinghagen and Zühlke-Robinet 2001).

In a fast-changing environment, a regulation-driven system can survive only if it is capable of reform. The reform of vocational training, especially the introduction of joint basic training for construction workers of different trades, and the reform of the Handicrafts Code have allowed the technical and organizational changes of the past decades to be not merely passively practiced, but rather actively co-arranged. It has proved beneficial in this context that the construction trade union was not organized according to trades and that it did not stick to traditional job demarcations, as was the case in the United States and the United Kingdom.

However, the change experienced over the past years called the entire system into question at a more basic level. Due to the internationalization of construction labor markets and the use of seconded construction workers paid at their domestic rates, wages once again became a parameter of competition. At the same time, the State withdrew from its construction labor market policy, the market started to decline, and German reunification led to a split labor market within the country itself. The social partners succeeded in carrying through a minimum wage rate for posted workers. They laid the foundation for international regulations being established by means of agreements concluded with the social funds of neighboring countries, and they have agreed upon new joint benefits, such as the bad-weather allowance. Equal in importance to these substantial arrangements was the stabilization of the increasingly fragile employers' associations. Here, the trade unions had to accept concession bargaining, giving back some previously won gains.

So far, the German construction labor market has not reached a new equilibrium. The differences between West and Eastern Germany will persist for a long time. In practice, the minimum wage rates are often not paid. It remains to be seen if the new provisions geared to enforce compliance with laws and collective agreements will have the desired effect. Upon EU expansion to Eastern Europe, the internationalization of markets will further increase. If the cohesion within the employers' camp continues to decline, collective bargaining will become more intricate. However, the social partners have caught an important tailwind from the new federal government over the past few years. The declaration of agreements on the minimum wage rate as generally binding was facilitated. The pressure exerted by the German government was key to achieving the situation where the future EU member states (Poland, Hungary, and the Czech Republic, among others) will be entitled to neither free movement of workers nor freedom of services for an extended transition period. The social partners will thus have some years of breathing room to prepare for a further linkage of the European construction labor markets.

NOTES

1 This chapter is based on the findings of a research project funded by the Hans Böckler Foundation and the Institute for Work and Technology. For the results of this research project, see Bosch and Zühlke-Robinet 2000.
2 It was a prerequisite of membership in the EU that national debt was reduced to 3 per cent of the GDP.
3 At the request of one of the parties to collective bargaining, the Federal Ministry of Labor and Social Affairs can, under certain conditions, in accordance with Paragraph 5 of the Collective Bargaining Act of 1949, declare a collective agreement to be generally binding, provided that the agreement of a collective bargaining committee made up of three representatives of the employers' and employees' central representative associations can be obtained. The following are the conditions that must be fulfilled: (1) employers who are members of the employers' association and are therefore covered by the agreement must have a workforce of at least 50 per cent of the respective industry; and (2) the declaration of general applicability must be in the public interest.
4 The costs rose because of both the increases in training allowances and the shorter time spent by apprentices in firms.

REFERENCES

Artus, I., Schmidt, R., and Sterkel, G. (1998) *Brüchige Tarifrealität: Tarifgestaltungspraxis in ostdeutschen Betrieben der Metall-, Bau- und Chemieindustrie*, research report, Düsseldorf: Hans-Böckler-Stiftung.

Bosch, G. and Zühlke-Robinet, K. (2000) *Der Bauarbeitsmarkt: Soziologie und Ökonomie einer Branche*, Frankfurt/New York: Campus.

Erlinghagen, M. and Zühlke-Robinet, K. (2001) "Branchenwechsel im Bauhauptgewerbe: Eine Analyse der IAB-Beschäftigtenstichprobe für die Jahre 1980 bis 1995," *Mitteilungen aus der Arbeitsmarkt- und Berufsforschung* 34 (2): 165–81.

Fürstenberg, F. (1988) "Kooperative Tarifpolitik", in F. Gamillscheg, B. Rüthers, and E. Stahlhacke (eds), *Sozialpartnerschaft in der Bewährung*, Festschrift für Karl Molitor zum 60. Geburtstag, Munich: Beck.

Hauptverband der Deutschen Bauindustrie e.V. (2001) *Baukonjunktur, Stand und Prognose, Herbst 2000*, Berlin: Hauptverband der Deutschen Bauindustrie e.V.

Industriegewerkschaft Bauen-Agrar-Umwelt (IG BAU) (2001) *Wirtschaftsdaten für Bauleute in Text und Bild*, Frankfurt: IG BAU Bundesvorstand.

Mavromaras, K. and Rudolph, H. (1995) "'Recalls': Wiederbeschäftigung im alten Betrieb," *Mitteilungen aus der Arbeitsmarkt- und Berufsforschung* 28 (2): 171–94.

Rußig, V., Deutsch, S., and Spillner, A. (1996) *Branchenbild Bauwirtschaft: Entwicklung und Lage des Baugewerbes sowie Einflußgrößen und Perspektiven der Bautätigkeit in Deutschland*, Berlin/Munich: Duncker & Humblot.

Schäfer, R. and Wahse, J. (1998) *Ergebnisse der zweiten Welle des IAB-Betriebspanels Ost 1997*, in IAB-Werkstattbericht No. 4 (20 May 1998), Nuremberg: Institut für Arbeitsmarkt- und Berufsforschung.

Statistisches Bundesamt (2001) *Fachserie 4 Produzierendes Gewerbe*. Reihe 5.1 Beschäftigung und Umsatz der Betriebe des Baugewerbes, various volumes, Stuttgart: Metzler-Poeschel.

Streeck, W., Hilbert, J., van Kevelaer, K.-H., Maier, F., and Weber, H. (1987) *Steuerung und Regulierung der beruflichen Bildung: Die Rolle der Sozialpartner in der Ausbildung und beruflichen Weiterbildung in der Bundesrepublik Deutschland*, Berlin: Edition Sigma.

Syben, G. (1999) *Die Baustelle der Bauwirtschaft: Unternehmensentwicklung und Arbeitskräftepolitik auf dem Weg ins 21. Jahrtausend*, Berlin: Edition Sigma.

Weitbrecht, H. (1969) *Legitimität und Effektivität der Tarifautonomie*, Berlin/Munich: Duncker & Humblot.

4 Denmark

Searching for innovation

Nikolaj Lubanski

INTRODUCTION

The Danish construction industry is more highly regulated than that of all countries in this study, with the exception of Germany and the Netherlands. However, regulation in Denmark, unlike Germany and the Netherlands, comes primarily through collective bargaining and the related standards developed by employers' and workers' organizations. The role of the State is active in the sense of involving the social partners but passive in the sense that state authorities do not set specific rules and regulations. The overall picture is that Danish construction has followed the high road of competition, combining strong private or voluntary regulation with high labor costs and high productivity levels.

The Danish system of voluntary regulation has been quite successful in meeting the construction needs of the Danish economy, establishing and insuring product quality, training both professional and craft workers in construction, and providing unemployment and disability insurance. Industrial organization has been relatively stable, despite the natural fluctuations in construction demand, and the wages of construction workers remain attractive within the Danish context.

Due to centralized collective bargaining, the wages across construction occupations are relatively even. The success of the Danish system is dramatized by the fact that it is common, customary, and accepted that Danish construction firms will guarantee their work for a significant amount of time after the completion of construction. The Danish model is therefore an example of construction stability through cooperation, with limited state involvement, primarily in the regulation of bidding procedures.

However, this system of centralized collective bargaining and cooperation among social partners is confronting a new challenge. By committing itself to the high-wage, high-skill, high-quality road, even in the face of increased international competition, the Danish construction industry must also promote ever-growing productivity. Fortunately, a precondition for this has long been in place, in the form of the Danish vocational education system, which provides to construction craft workers a combination of classroom theoretical learning and workplace practical learning. Professionals are trained in technical universities funded by the national tax system. However, despite this well-functioning system of construction training, overall productivity in construction has been stagnating in recent years, leading to a crisis.

Concerned parties – consumers of construction services, construction contractors, unions and the government – have addressed the issue of stagnating productivity by looking at potential reforms of work organization. Currently, work is organized along occupational lines with both workers and subcontractors forming themselves into occupational specialties (e.g., electricians and electrical contractors), as in the United States. Proposed reforms suggest a reorganization of construction with a greater focus on product rather than occupation. Some proposals include partnerships – long-term cooperation between groups of specialized firms that present the consumer with the full range of building specialties. Thus, contractors would not come together simply by accident at a given building site, but would rather form a group with longstanding experience in cooperating with each other. Another potential reform is the creation of mixed-occupation contractors who bring to the job site workers with a range of occupations. Both these reforms look to adopt aspects of workplace management from manufacturing.

In addition, forms of worker–management cooperation are being considered. In the construction context, worker–management cooperation involves not only unions and contractors but also the consumer of construction services (i.e., the owner). New bidding procedures are being considered where the entire building process is planned from scratch, and the work crews are composed on the basis of product plans jointly developed by owners, architects, engineers, contractors, and workers' organizations. In contrast to the United States' "design-build" construction, in the Danish case, wider participation is envisioned, including not only contractors but also unions in the planning stages of construction. The success of these reforms will require stability and cooperation, conditions that are often notably absent in the construction industry of other countries, particularly less regulated ones. The tradition of private, voluntary self-regulation in Danish construction provides the baseline of stability and the tradition of cooperation that make these reforms possible. The government, trade unions, and employers' organizations are now in the process of negotiating the implementation of these or similar reforms. Danish construction seeks to remain competitive in a more globalized environment by emphasizing its traditional strength – not specific regulations but a tradition of developing mutually beneficial regulations through voluntary self-organization and collective bargaining.

The success of reforms remains to be seen. Pressure to change is coming primarily from competition with Swedish construction firms. These firms are entering Danish markets primarily to maintain and exploit economies of scale rather than to compete based on low-wage, low-skilled labor. Consequently, it has been relatively easy to integrate Swedish construction workers into the matrix of unions and collective bargaining in Denmark. Responding to Swedish competition and the development of large projects that fall under EU rules for bid tenders, there has been a hollowing out of medium-sized construction contractors in Denmark. A handful of very large contractors has emerged alongside a very large number of small contractors. Thus far, Denmark's regulatory system has insulated the local construction economy from the sharpest competitive pressures from low-wage, low-skilled contractors from other parts of the EU.

Until now, the challenge of sustaining a high-wage, high-skilled, quality-driven construction industry has been met through a range of specific measures that have left the general structure of industrial relations relatively untouched by the winds of internationalization. The future challenge is whether Denmark's system of voluntary self-regulation and centralized collective bargaining will prove sufficiently prescient and flexible to respond to the continuing challenges posed by international competition.

In this chapter, the background for this challenge will be further explained and placed in the framework of the general development of the Danish construction sector over the last four decades. The second section will outline the economic evolution of the sector. This is followed by an analysis in the third section of the development in employment and the related company structure. In the fourth and fifth sections, the characteristics of the product market and of the labor market are investigated with regard to identifying the regulatory framework. On this basis, it is possible to understand the context of recent changes in the sector and to present challenges in the sixth section. Finally, the last section presents conclusions.

ECONOMIC EVOLUTION

This section deals with the development and economic evolution of the construction sector in Denmark from 1966 and provides an overview of the industry's position in the Danish economy as a whole.

The development of the construction industry in Denmark has been characterized by three main tendencies. First, the total production in the sector has increased but with significant variations among subsectors. The construction of new buildings has fallen remarkably behind the development in civil engineering[1] and repair/maintenance activities. Second, the structure of the sector has changed. Medium-sized companies are disappearing, giving way to a new structure with a few large companies at the top and dense undergrowth of small, specialized firms. Third, price competition is high and increasing. Although price competition has always been significant, the structural changes have intensified the battle for position in the market, leading to low rates of earnings even in periods of high activity.

Figure 4.1 indicates that the development in production of the Danish construction sector from 1966 to 1996 as a total has increased. The repair/maintenance curve shows a production that has more than doubled, and civil engineering has had a similar development. However, the construction of new

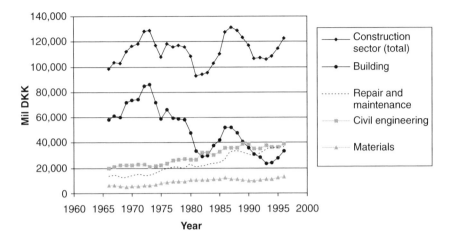

Figure 4.1 Production in the Danish construction sector, total and by subsector 1966–96
(1995 prices).

Source: Danmarks Statistik 1997.

buildings shows a marked decrease in production. The production of materials
plays a more or less insignificant part in the total production development.

The total production value in the Danish construction sector is defined as
the value of market and non-market activities. The total production value for
construction was 114.6 billion Danish kroner (DDK) in 1995, compared with
106.2 billion DDK in 1993, corresponding to an increase of 7.9 per cent.

The construction sector is often one of the sectors with most irregularity in
terms of business cycles. With the decrease in demand for construction of new
buildings in the 1970s and 1980s in Denmark, the production of the sector fell.
This is also indicated by the construction sector's share of the total Danish gross
factor income, which in the period 1970–2 was nearly 12 per cent, but in the
period 1989–91 had fallen to only 5.5 per cent. An important turning point
occurred in the years around 1973 and 1979, when the oil crises had a major
impact on the Danish economy in general and on the level of activity in the
construction industry in particular. While the construction sector now has a less
dominant role in the national economy, its contribution to the fluctuations and
the total activity in the Danish economy is still important.

After a boom that finished in the late 1980s, the construction sector experienced approximately 5 years with a lower level of activity. In spite of strong
growth since the early 1990s, the level of activity in 1996 was still lower than in
1988. A sign of the sector's persistent dependency on the overall development in
the Danish economy is that the growth through the last decade has mainly taken
place in the building sector. Since the prices of privately owned residential
houses have been rising since 1993, and the prices for leases of office buildings
have gone up as well, new buildings have been able to compete with the prices of
the existing housing market.

Throughout the construction sector, medium-sized enterprises (50–200 employees) are disappearing (Danmarks Statistik 1997a: 6). The concentration trends develop because many companies have not been consolidated after the period of economic recession up to 1993 and they are therefore vulnerable to price competition. This helps also to explain the generally low level of earnings of Danish contracting companies (Entreprenørforeningen 1997). The explanation offered for the low earnings has, in recent years, been that a number of building projects were still under way that had been contracted at too-low prices during the years of crisis in the early 1990s. As the years pass and the recovery continues, it becomes increasingly difficult to maintain this explanation. The more plausible explanation is that medium-sized enterprises are being squeezed by the intense price competition that has resulted from the fight among a small number of big companies for the leading position in the Danish construction market.

EMPLOYMENT

Figure 4.2 shows the development in employment from 1966 to 1996. The top curve represents the total number of employed in the Danish construction sector over the period. The three curves below represent the three main subsectors in the Danish construction sector: building (i.e., construction of new buildings), repair/maintenance, and civil engineering.

The total number of workers in the construction sector fell from 205,000 in 1966 to 150,000 in 1996. Initially, in the second half of the 1960s, the total grew. However, by the beginning of the 1970s, this number started to fall, with the most marked decreases after 1973 and 1979. Again, after 1985, total employment increased, from 157,000 to 173,000 only 2 years later. However,

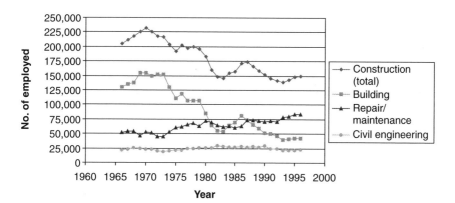

Figure 4.2 Employment in the Danish construction sector, by subsector, 1966–96.
Source: Danmarks Statistik 1997b.
Note: The number of employed is a total that includes hourly workers, salaried workers, foremen, and others.

after 1987, it decreased once again, to a low of 139,000 in 1993–4, bouncing back up slightly thereafter. These figures clearly indicate that over the long run Danish construction employment is decreasing and highly influenced by cyclical effects.

The rising level of activity in the sector since 1993 is reflected in the number of employees. Approximately 4.3 per cent of the total labor force in Denmark in 1995 was employed in the construction sector, not including white-collar workers. This corresponds to an annual average of 112,870 employees. In 1997, this figure had grown to 115,392 employees. The number of employees continues to be somewhat under the 1988 level with an average level of 129,522 employees, but is an increase compared with 1993, when there were only 101,701 employees in this sector.

In contrast, the number of white-collar workers employed in this sector has been rather stable with 26,416 in 1995 and with 26,448 employees in 1998. In line with international trends, while the number of owners fell by 28.6 per cent from 1995 to 1998, the number of employed white-collar employees grew by 9.1 per cent. The Danish sector has not witnessed a growing number of self-employed workers in the sector. Compared to the situation in the United Kingdom, for instance, the proportion of self-employed is very low.

Two-thirds of all employees in the sector are employed in small businesses with fewer than 50 employees. Less than one-fourth is employed in companies with over 100 employees. This indicates both that there are very few large companies in Denmark and that they are quite small in an international context. From a total of more than 17,000 companies in the sector, fewer than 100 are employing more than 100 employees. The majority of firms have fewer than 5 employees.

REGULATIONS IN THE PRODUCT MARKET

The regulation of the Danish construction product market stems from many sources. Seen from an international perspective, a relatively small number of regulations are based on laws. Instead, several institutions and organizations have influence on different parts of the regulations. At first glance, the sector might seem unregulated in a number of areas; for example, there is in principle no registration of contractors in Denmark. However, as we shall see, a closer look reveals the pattern of regulation.

Institutional and legal framework

A significant force behind the regulation of the product market is the government. The most prominent government regulatory bodies are the following:

- The Ministry for Urban and Building Affairs, through its important council, the National Housing and Building Agency, is responsible for most regulations relating directly to the building process (e.g., quality matters, building regulations, subsidies, and the approval of materials), and to housing concerns (e.g., the Law on Renting, housing regulations, and town renewal).

- The Ministry of Environment is responsible for the strategic national physical planning, including regional planning, energy policy and conservation, and environmental protection. During the last decades the role of this ministry has gained more importance, not the least through the increased focus on environmental issues.

- The Ministry of Traffic is responsible for overall transportation infrastructure. Civil engineering and maintenance activities are therefore governed by regulation from this ministry.

Next to these three ministries, also the Ministry of Education and Ministry of Industry are influencing certain fields of activities in the sector. Furthermore, local authorities play an important role through their responsibility for local planning and building permits. Applications for a building license to start construction must be sent to the local building authorities. Local governments also ensure that building regulations are respected. After the construction work is completed, they are the ones who grant permission to use or occupy the building.

The Danish Housing Act of 1966 set out the building regulations and the system of approval for new building materials and construction. However, most Danish building regulations are not based on legislation. Rather, they have been established through tradition by contracts between the parties and by the administrative regulations of the authorities.

Segmentation of the industry and tendering

Three main contracting methods are used in Denmark: turnkey contracting, main contracting, and specialist trade contracting (Atkins International 1993).

At present turnkey contracting is the most common. Under this contracting form, the client – i.e., the owner – directly contracts with only one contracting party (the turnkey contractor) and not with a number of individual contractors, suppliers, etc. The turnkey contractor undertakes both to design the project, with the assistance of an architect and engineering consultant, and to complete it. The tender of the turnkey contractor is submitted on the basis of not a fully detailed project, but rather of a concept specifying the client's requirements. This contracting form is mostly used for the construction of standardized buildings. Public authorities are also increasingly using this method, because they are less involved in controlling the details of design and construction.

In main contracting, an invitation to the main contractor to tender comes from a consulting engineering firm on behalf of the client. The consulting engineering firm has worked out a detailed project for the client and invited tenders on the client's behalf.

Specialist trade contracting is very important in Denmark. A well-developed subcontracting system has led to the large number of small enterprises in the

Danish construction industry. Specialization is an important issue for small and medium-sized enterprises (SMEs) in order to win business. Specialist contractors are taking a larger proportion of the construction work of the general contractors, while the latter increasingly act as managers only. Specialist subcontractors normally operate regionally rather than nationwide.

In 1996, the Tender Act was adopted. Its relatively late introduction was in response to the increasing use of restrictive trading agreements in the sector. The Act has formed the basis for the rules regarding the submission of tenders and the purchaser's acceptance of tenders of bids of two types: (1) public tenders, where any company may submit a tender; and (2) contracting by tender, where only a limited group of selected firms is invited to submit bids for a specific task. The Act on tenders has served several purposes in changing the tendering process. First, it has restricted the purchaser's right to call for private bids. Second, it has created a process for publicizing the submitted bids. Third, it has imposed restrictions on the purchaser's further negotiations beyond the tendering round. The Act has been successful in its goal of making the competition in the construction sector more transparent.

Control of contractors and quality assurance

As stated above, the legislative foundation of the Danish construction sector might seem shallow compared to some other European countries. Instead, the industry is partly self-regulating. For example, in principle, there is no government registration or authorization of contractors. "Contractor" is not a protected title in Denmark. Anyone may start up a construction firm, as long as the regulations regarding skill and quality are met. Over the last few decades, it has been frequently debated whether some sort of general authorization should be introduced; however, each time the different interest organizations have rejected it as hindering free competition. Only artisans such as plumbers, sewer contractors, chimneysweepers, and electricians (or companies offering one of these services) must become authorized. A form of private registration is instead taking place at the Danish Contractors Association and the Federation of Small and Medium-Sized Enterprises in their role as employers' organizations.

Likewise, there is no legislation in Denmark governing entering into a contract. However, the General Conditions for Works and Supplies, passed in 1972 and revised in 1992 (*Almindelige betingelser for rådgivning og bistand*, or AB 92), has generally been accepted as the basis for entering into a contract and for the details that the contract should cover. AB 92 covers the supply of materials under the contract and the materials supplied directly to the purchaser. It includes ground rules for the invitation to tender, the tender itself, and the subsequent entering into contract. In setting up these conditions, the sector's organizations were highly influential, underscoring the industry's partly self-regulating status.

As part of AB 92, the liability issue and regulations on guarantees and insurance are also covered. After delivering work, the contractor is liable for 5 years – 20 years for civil engineering work – for any failure attributable to that work. Thus, the contractor is responsible for defects caused by employees' mistakes. Purchasers must check materials on receipt, and any complaints regarding visible defects must be made immediately. Approximately 5 years after handing over, an inspection of the construction takes place. AB 92 includes regulations on project review, control plans, documentation, and tests to heighten quality awareness and assurance in construction. Concerning guarantees, the contractor must provide security for a contract of 15 per cent of the total contract sum upon acceptance. This sum is reduced to 10 per cent after one year and to 2 per cent for the remainder of the 5-year period.

Special bodies have been established to check that safety and quality requirements are met. Historically, these bodies specialize within their own fields of products and application, doing the testing, certification, and approval. In the industry, there is a general acceptance that freedom of method is essential for the evaluation of quality assurance systems in construction. International Organization for Standardization (ISO) standards are gaining further ground in the sector, and over the last decade the tendency has been to view quality assurance as part of the total building process rather than as a separate task providing competitive advantage. Nevertheless, it is still perceived as an infringement of freedom in Denmark to have the production method strongly regulated by manufacturing norms and quality requirements.

REGULATIONS IN THE LABOR MARKET

The Danish construction labor market is highly regulated, not least by its strong interest organizations. The sector carries many of the characteristics of the so-called Nordic labor market model, as a largely voluntary system with strong employers' organizations and trade unions, high coverage rates, and limited potential for state interference.

Industrial relations

The most important features of the Danish industrial relations (IR) system can be summed up as follows: (a) a high rate of organization on both the employers' and employees' sides; (b) a strongly centralized collective bargaining system; (c) close relations between the State and the social partners; and (d) a basic form of regulation with an emphasis upon collective agreements concluded between the social partners rather than legislation (Strøby Jensen 1995).

With the so-called September Compromise from 1899, two principles were firmly established: (1) the employers' management prerogative to manage and control the work process; and (2) the right of trade unions to organize and negotiate on behalf of their members. Since then, these two principles have been the main pillars of the Danish IR system. One of the consequences of the September Compromise has been a normative and political orientation in the direction of concluding agreements rather than enforcing legislation. Both the trade unions and the employers' organizations have seen it as their task to reach some form of agreement with their counterpart, rather than leaving differences to be resolved through regulation by the State. This basic orientation towards collective bargaining has, at the same time, contributed to the organizational development of the social partners and has consolidated their decisive role in the regulation of the labor market.

In Denmark, labor market rights have primarily been obtained through membership in a trade union, which concludes collective agreements, and not by government legislation. This has contributed to ensuring the high level of union membership in Denmark, providing incentive to the individual employee to join. Furthermore, the employers' organizations at an early stage accepted the employees' right to organize and to collective bargaining, and Danish enterprises continue to be rather positive towards organization of the employees. This means that the development of the Danish labor market has not been characterized by systematic attempts of undermining the *raison d'être* of the trade unions as have been seen, for instance, in France and Great Britain (Ferner and Hyman 1992).

The employers' associations have contributed similarly to maintaining the characteristic features of the IR system. First, the employers have also been generally in favor of regulation of labor market conditions by means of collective agreements rather than legislation. Second, the Danish Employers Confederation (*Dansk Arbejdsgiverforening*, or DA) has had an interest in a centralization of the bargaining system. Since the majority of Danish enterprises are small or medium-sized, the employers strategically tried to centralize the negotiations quite up to the end of the 1970s, to prevent the trade unions from playing the many SMEs against each other in the bargaining situation (Due, Madsen and Strøby Jensen 1993).

In the construction sector, the centralization of the organizations, mainly on the employer side in DA and later in the Confederation of Danish Trade Unions (*Landsforeningen i Danmark*, or LO), meant that there are fewer parties and fewer agreements today than earlier. The employer side is represented in the negotiations by one of the organizations in the so-called "square" of the building sector: the National Association of Electrical Contractors (*Elinstallatørenes Landsforening*, or ELFO), the National Association of Employers in the field of Water, Heating and Sanitary Installations (*Dansk VVS Installatør Forening*), the Danish Construction Association (*Danske Entreprenører*), and Employers in the Building Sector (*Byggeriets Arbejdsgivere*, or BYG). A division of labor has developed where the Danish Construction Association represents mainly the contractors, and BYG represents mainly the smaller craft enterprises, this difference representation often provides the dividing line between main contractors and subcontractors. The Danish Construction Association represents the building contractors and primarily the big companies.

On the employee side, the employees are represented through the seven unions that are organized under LO in the Building, Civil Engineering, and Wood Industries cartel (*Bygge-, Anlægs- og Trækartellet*, or BAT cartel) with the

Unions of General Workers in Denmark (*Specialarbejderforbundet i Danmark*, or SiD) as the largest followed by the recently merged union Wood–Industry–Building (*Træ–Industri–Byg*, or TIB). BAT is a cooperation cartel, which does not have the right to negotiate the renewal of collective agreements; the individual unions carry out these negotiations. However, common negotiations are taking place with BYG through BAT. In 1997, the BAT cartel had 159,000 members; the SiD, 67,786 members; and TIB, 55,830 members. The rate of union membership among the employees in the sector is very high. Most estimates are between 90 and 95 per cent (EFBWW 1995).

Collective agreements

The national sector agreements between national trade unions and employers' organizations cover most of the Danish labor market and thus play a central role in the Danish IR system (Due, Madsen and Strøby Jensen 1993). The regulation of the construction sector largely takes place at the sector level, through collective agreements between the social partners. To a large extent, these agreements have the power to set the standards in each sector, because of the high rate of organization on both sides of the table. For example, the vast majority of the employees in the construction sector are covered by a collective agreement. These agreements cover wages and working hours, as well as a whole range of other conditions, such as pension schemes, safety matters, and the obligations of shop stewards.

The centralized agreement system has also led to a high degree of homogeneity in the pay and working conditions of construction workers in Denmark compared with other EU countries (Ferner and Hyman 1992). There is only a moderate wage spread among the employees, for example, and historically there have been only modest differences in the weekly working hours among different groups of employees.

The local trade union branches ensure that the agreements are being complied with and act as a link between the union and its members through, for instance, collecting information on pay conditions and training shop stewards. The degree of compliance with the agreements is very high in the Danish system. Denmark has a well-functioning system for settlement of industrial disputes, which takes disputes out of the workplace. Here again, the local branches of the unions are the backbone of the system. Surprisingly, there are no prevailing wage laws. The Danish unions themselves are sufficiently strong to ensure that the rates and benefits of the collective agreement are paid by the employer, using strikes, blockades of sites, etc., if the normal procedures for settling industrial disputes fail.

The agreement in the construction sector between SiD and the Danish Construction Association is an example of the national agreements covering most employees. The two dominant chapters in the agreement deal with working hours and wages. The changes in these two fields reflect the general changes taking place in the collective agreements during the last two decades.

Significant changes have been introduced in relation to working time. The workweek decreased from 40 to 37 hours from December 1986 to March 1990 and has since been held at that level. However, while the frame of the agreement is settled at the national level, the ability of companies to schedule working time within the frame has increased. Formerly, there were strict limitations on the companies' ability to arrange work hours to meet their needs, but within the last decade this decision-making power has been decentralized to the company level. It is now possible to create agreements at the local level about overwork, weekend work, shifting working weeks, and time off.

Similar changes have been introduced in the wage field. In 1991, the Danish Construction Association and SiD agreed that the so-called "normal pay" system applicable to non-permanent workplaces and the minimum wage system for permanent workplaces should be merged into a common wage system based on minimum pay. The minimum pay system is the most decentralized of the three wage systems. While the minimum pay rate is fixed in the collective agreement, most employees are paid considerably higher than the actual minimum wage, and they can obtain wage increases during the term of the agreement only by means of local negotiations. The supplements to the minimum pay rate are often based on piecework rates that are also fixed in the collective agreement. Tariffs determine the functions to be performed as piece-rate work and must be approved by the parties.

In spite of the present favorable times for the sector, wage development has been lower than that in other sectors, in contrast to earlier periods of economic recovery. There are a number of explanations for this. The most important is probably the general awareness among both enterprises and employees of the consequences of very big pay increases. In the earlier periods, the competitive spiral resulted in high levels of inflation, which did not benefit any of the parties. Today we see a shift in the direction of stability and job security.

Recruitment for performance of a special task is the most common form of recruitment among the workers in this sector. Individual workers are recruited to perform a specific task or project for a given period of time. Because of the uncertainty of this employment relationship for the worker, the unemployment insurance system has been structured so that the employee contributes directly to an unemployment fund, and the employer has no responsibility after the normal layoff procedures. Thus, workers receive unemployment compensation for periods of unemployment with a guaranteed basic income, and employers have a high degree of flexibility in hiring and firing workers for specific projects.

However, the trend in the last few decades is toward formation of work gangs, which are actually employed with the enterprise rather than on a project or a time-limited contract. Many such gangs stay with a single enterprise for a number of years, although they are formally employed at one project after another.

Individual enterprises typically have a core labor force that stays with the enterprise for many years with *de facto* permanent employment. The core labor force is characterized by such factors as a high level of qualifications, stability, and loyalty. According to figures from the sector's pension fund, out of the approximately 30,000 employed in this field, about 25 per cent have a long-term relationship with an individual enterprise.

Compared to other countries, such as the United States, the Danish system of collective bargaining is still very centrally coordinated in spite of various processes of decentralization. The latest changes have been initiated chiefly by the employers' organizations, motivated by a wish to create a broader framework for negotiations in the individual enterprises. However, there is no significant deregulation of the sector's IR system in Denmark, and the collective agreements remain the principal regulating factor on the construction labor market.

Vocational training

Most of the education system in Denmark is public, but, in relation to the adult education system and vocational training, the labor market organizations are very influential. Most workers in the sector are skilled, and the reproduction of the trained workforce is secured through educational funds. For professionals in the sector there is one way to become an architect and three possible ways to become an engineer, all related to public schools and universities.

Vocational education combines on-site training with formal education at different technical schools. The Danish vocational training system (*Voksen og Efteruddannelsessystemet*, or VET) can be described as a cultural bridge between the German dual apprenticeship system and the school-based models of the other Nordic countries, with more time spent on theoretical training in school than in Germany, and conversely far more practical in-company training than in Sweden (with 60–75 per cent in Denmark compared to around 15 per cent in Sweden). Vocational education in Denmark is centralized, and the Ministry of Education lays down all standards in the form of regulations. These regulations are agreed upon by the social partners, and then approved by the Ministry. Training generally lasts for about 4 years; significant degrees of flexibility were introduced after the latest reform of the system in 1999.

Vocational education encompasses the building trades, which are very specialized in Denmark. The main categories are bricklayers, carpenters/joiners, painters, plumbers, and electricians; except for the last category there are various specializations under each of these main categories. Following the latest reform, a basic program was established that the student completes before entering into the vocational program where an apprenticeship contract is signed with a company. The employer pays wages to the apprentice, including wages for the periods spent at school; these latter are reimbursed from the Employers' Reimbursement Fund.

Upon finishing vocational training the worker is certified as a journeyworker. Any journeyworker who is a manager can be a master craftsperson. As indicated before, there are no public licenses or certificates. The qualifications obtained from vocational training are recognized nationally and by employers and employees alike, as their own representatives participate in the development and implementation of the curricula and monitor the examination results.

The high degree of specialization in vocational training has been relatively successful in skill formation. The smaller and specialized building enterprises have taken advantage of being able to employ apprentices with a high level of qualifications in their respective fields of specialization. On the other hand, the specialization has also led to rather rigid demarcation lines between enterprises, and those do not always match the changes in production processes. Owing to these problems and to the fact that the construction sector still employs a relatively high number of formally non-skilled workers, a new generalist program has been introduced in the last decade. Under the program, the worker is given a broad contractor-like education, allowing experienced but non-skilled workers also to obtain an education in either civil engineering or construction.

Many education initiatives are under taken by the Educational Funds that have been established by the social partners. These funds are financed through contributions paid by employers per hour worked. The precise amount is stipulated in the collective agreements.

Other social funds

The collective agreements have also introduced pension plans during the last decade. Before the early 1990s, pension schemes were primarily for professionals and public employees; construction workers had state retirement pay only.[2] Through collective bargaining, the social partners have settled on a system with joint payment from the company (two-thirds) and the worker (one-third). From a modest start, the percentage of the total wage set aside for the pension funds has grown to a level more comparable to other groups in society.

In Denmark, all unemployed workers are entitled to unemployment benefits. However, members of an unemployment fund can get higher benefits. The individual worker pays the contributions to these funds, and the payments are subjected to tax deductions on a yearly basis. Since the unemployment benefits are relatively high, there are continual discussions about whether the economic incentive for the unemployed to search actively for new work is substantial enough. The level of the unemployment allowances also explains why the sector does not experience an even greater outflow from the sector in spite of the temporary nature of the production processes.

Traditionally, there are close links between being a member of a trade union and paying into an unemployment fund. The close relationship is often given as an explanation of the high rates of organization on the Danish labor market. Social security can also be obtained when an employee is ill, has been injured, or is expecting a child. Furthermore, regulations of allowances for workers when construction work cannot be carried out due to bad weather are set in the collective agreements.

CHANGES AND CHALLENGES

Seen on an international scale, the winds of change do not blow with the same strength in Denmark as in many other countries. The Danish construction industry has experienced neither a re-unification process as in Germany nor a unilateral attack on the collective bargaining system as in the United Kingdom. Nevertheless, some lasting changes have occurred during the last few decades, leaving the sector with a number of challenges to overcome if the high road of competition is to be continued.

The key terms in this regard are innovation and internationalization. To some extent these two terms and the processes they involve are of course related. Continuous innovation is important because otherwise the small and medium-sized Danish companies are in danger of being taken over by foreign multi-nationals. Likewise, the relatively high labor standards can be maintained only if productivity is rising. While these two processes represent genuine challenges of a newer date, the way they are met by the parties in the Danish construction sector seems more in line with earlier forms of adaptation. The joint efforts of the social partners to change the sector in line with the present challenges (i.e., to push for more innovation) cannot be seen as a break with the past. Even in periods with severe economical downturns, the social partners have tried to find common ground for the solution of the problems in the sector.

The search for innovation and further productivity gains

Productivity in the Danish construction sector is characterized by a slow growth rate – if there is any growth at all. For the building of new dwellings the national accounts show a stagnating and even decreasing productivity from the 1970s onwards. Analyses of time consumption in the building sector show a strong growth in productivity in the 1950s and 1960s, followed by a declining trend in the 1970s and 1980s. The overall picture is of a sector that by and large has had a stagnating productivity since the end of the 1960s.[3]

The main reason for this lack of productivity growth is "the rise and fall" of large, standardized industrialized building, which results in a streamlining of the work carried out (AE 1997). This streamlining happened for two reasons. First, the delivery and assembling of elements required a high degree of planning. Second, the historical problem of coordinating the craftwork among subcontract-ors divided by trade became smaller. In the 1970s, industrialized building became less significant, and smaller series of production became more predominant in the two following decades. The industrialization of the building industry is one aspect of the question of productivity.[4] The other has to do with the coordina-tion of the work process on the building sites.

The challenge today in the Danish construction sector is to increase produc-tivity by organizing the building process more rationally. This aspect is accentu-ated as the present demand for buildings is focused on smaller production series that appear non-standardized. Also, the share of repair and maintenance in the sector is growing and can be expected to increase in the future. What has been suggested is a reorganization of the work process on the building sites, shifting from organization along trade divisions of different specializations to organiza-tion on the basis of a so-called product orientation. The idea is that building or reparation work should not be done as a project carried out by perhaps three companies working in parallel and divided by trade. Instead, it should be a project carried out across these companies in common. These changes in the organization of the work process demand substantial planning efforts because the work has to be coordinated between and across a number of different companies.

Over the past 30–40 years, the question has regularly been raised whether the traditional organization of the construction sector by trade sets up a number of barriers that reduce the efficiency in the building process. The focus is on whether it has hampered the development of a broader cooperation culture. The cooperative relations have often been characterized by disputes, the use of economic means of compulsion, sub-optimization, claims, and counterclaims. As described by the Development Council of the Building Sector (*Byggeriets Udviklingsråd*, or BUR), the resources have been tied up in traditional processes, rather than focusing on an optimization of planning and cooperation processes with a view to higher common profits. In connection with their "New Forms of Project Cooperation" in 1998, BUR drew the conclusion that the organization and the rigid occupational demarcations explain why the productivity in the industry has been lagging considerably behind the productivity development in other sectors.

Based on the many indicators, the vision of both the labor market organizations and the government has been to give new impetus to innovative thinking in the construction sector. A new industrialization of the building process should take place so that it will become product-oriented to a higher degree rather than split up along occupational lines.

The productivity and the cooperative relations in the building sector came into focus with the "Building Policy Action Plan" published by the Ministry of Urban and Housing Affairs (*Boligministeriet*) in April 1998. The Government wished to promote the quality and productivity within the building sector. This Plan took up the discussion about the special structure of the building industry. It aimed to improve the productivity in the building industry through the development of new and more flexible cooperation forms, which might contribute to improving the organization and management of the building process. The Plan proposed a number of possible cooperation models that it proposed should be tested in the non-profit housing sector, in public building projects, and in the private building sector in the coming years:

- *Partnerships* – The interested bidders sit down together and plan the building project so that it suits the strength of their individual firms. A long-term cooperation between groups of firms for two or three building projects is set up under the incentive agreement, which will be for the benefit of both the builder and the firms.
- *Horizontal industrialization* – This involves better cooperation and organization of the building site, among other things, by the use of mixed gangs, logistics, and planning principles from the manufacturing industry.

- *Intermediate bids* – The builder submits the tender on the basis of the program drawn up. Several bidders submit a basic proposal according to the program designed.

From the late 1990s, the focus has mainly been on the potential in a reorganization of the work processes in the building industry. After having been based for many years on the traditional gang structure, the work organization in the construction sector finally seems to be in process of change. There is a growing awareness in the sector that productivity gains can be obtained only by abandoning the rather rigid demarcation lines and by increasing the emphasis on a coordination of the entire building process.

A report from the Industrial Council of the Trade Union Movement (*Arbejderbevægelsens Erhvervsråd*, or AE) from November 1997 also deals with the development in work and organization forms in the building sector. The report recommended the introduction of a new type of tender that is more in style with the partnerships from the Building Policy Action Plan. The idea of partnership has kindled a keen interest and has been described as very promising. The tender experiments will take a form where the entire building process is planned from scratch and the gangs are composed on the basis of products rather than occupation, as has been the tradition for many years. It is then up to the companies to either form a multi-functional crew within the company or enter into a cooperative arrangement with other firms.

The plans for the development of the sector's organization of working processes again indicate the way severe challenges in the sector are faced. In spite of disputes between the government, the trade unions, and employers' organizations about work organization, they nonetheless negotiate with the aim of finding common solutions to the problem of innovation.

Internationalization and the issue of posted workers

The internationalization of the construction sector has not had as strong an impact in Denmark as in other countries in Europe (Lubanski 1999). The Danish market has, to an increasing degree, been opened up to foreign companies over the past decade, but the internationalization process has mainly taken place through Swedish (and, to a limited degree, French and Finnish) multinationals taking over Danish companies.

The requirement to open up tasks of a certain size to EU tenders means that foreign companies may bid for contracts in Denmark. We have also seen a process of mergers, mainly to match the increased international competition; for example, two of the top six companies have merged to create the largest contracting company in Denmark.

Thus, the development within the sector is no longer exclusively conditioned by domestic factors. This is seen most clearly in the big infrastructure projects. Within the EU, the implementation of big infrastructure programs has been seen as useful for developing a single market and supporting the economic development in the member states. At the same time, the trans-European infrastructure projects are of such magnitude that they call for the creation of consortiums with actors from different member states. The construction of the Öresund Bridge between Denmark and Sweden is a fine example of this. The project has been supported by the EU, and the EU rules for tenders have been followed with the result that a number of foreign companies have been involved.

The internationalization of the Danish construction sector has been of limited importance to industrial relations in general and to labor market standards in particular. An important stabilizing factor for Danish labor market relations is the highly organized labor market, both on the employer and on the employee side. This IR system also applies to foreign companies operating within Denmark. As early as 1993, foreign companies operating in Denmark concluded an agreement between DA and LO on how to tackle the problem of the posting of workers. This agreement requires the companies to observe the rules laid down by collective agreements, and DA will not object to union actions organized with a view to achieving this aim.

Control of compliance with Danish collective agreements in foreign companies is largely left to the social partners themselves. This is where the high rates of organization play an important role (Clegg 1976). Within the construction sector, it is the local branches of the different trade unions that are responsible for ensuring the coverage of employers within their occupational/geographical field. These local branches have been successful in obtaining this goal. The most spectacular case to date was the building of the National Library in the capital, where an Italian company with posted workers was accused of not following the Danish regulations. After a long blockade of the site, work continued only after the employer had paid compensation to the workers, bringing the foreign workers up to the Danish salary level.

The regulation of posted workers also rests on cross-national agreements between trade unions. There is cooperation under the auspices of the Nordic Building and Woodworkers Federation (*Nordisk Bygge og Træ Føderation*, or NBTF) with regard to cross-border problems such as the posting of workers and commuters. This cooperation received further emphasis following the expansion of the EU to include Sweden and Finland, and it led in 1996 to an agreement that sets out guidelines for the handling of employees who work across borders.

The main principle of the agreement is that trade union membership for a commuting or posted worker is transferred to the local union in the host country. The unions undertake to ensure that the transfer takes place and that the worker can be accepted into the local branch without paying additional fees, and that reasonable service can be obtained in the host country. When their members work in another Nordic country, the unions must also help them comply with national legislation and/or collective bargaining in that country. If disputes arise regarding conditions of pay and employment, the union in the host country handles them.

In 1998, this NBTF agreement was supplemented by an agreement between the BAT cartel in Denmark and the main construction union in Sweden,

Byggnads. The purpose of the agreement is to ensure the fewest possible organizational obstacles and complications in connection with work in the neighboring country. This agreement is based on the NBTF agreement but clarifies it and adds further points.

To explain the limited influence of internationalization processes on the construction sector's IR system, two further points should be observed. First, it is important that the influx of foreign labor has been relatively limited. Denmark has not at all experienced the same number of posted and/or illegal workers, as has Germany, for example. The Swedish, French, and Finnish multinationals are producing mainly through subsidiaries that use Danish labor, and are only to a limited extent bringing their own labor forces with them. The relatively smaller construction market, the longer distance to countries with a labor reserve, and the language barrier can be considered as supplementary explanations. Second, cross-national financial investment into the Danish sector is primarily coming from the other Nordic countries, which are not mainly internationalizing in order to compete on the basis of lower pay and working conditions. They are rather going international to achieve greater economies of scale and to match other multinationals on the European market.

For instance, Swedish construction companies have engaged in activities in Denmark chiefly because of the poorer business conditions in Sweden. However, they also are willing to invest to gain a foothold on the north European market as such for the long term, especially since a number of companies are currently trying to build-up their activities in the market. They are interested in presenting themselves as possible alternatives to Danish enterprises. This again means that the process of internationalization takes places in a regulated fashion, being accommodated by the existing system instead of being a major threat to it.

CONCLUSION

When viewed over the last four decades, some aspects of the Danish construction sector have changed significantly, whereas others are characterized by fewer changes. On the one hand, huge structural changes have taken place, leaving the sector with a less significant role to play in the national economy and with a polarized industry structure. On the other hand, the division of labor between the contractors and subcontractors to a large extent remains the same, and some of the problems – for example, concerning lack of productivity – are still present.

The Danish construction sector is an example of voluntary regulation through collective bargaining and self-organization on the part of employers and workers with relatively minimal government regulation. This system has worked well in creating a high-wage, high-skill construction industry that builds quality products that are insured for a long time compared to other countries. Internationalization has put pressure on this system, but the social partners have maintained a high-wage, high-skill system through agreements that uphold Danish standards even for posted workers and workers employed by foreign subsidiaries. Nevertheless, this pressure has revealed the problems of stagnating productivity despite an effective training system. Thus, owners, contractors, workers, and the government have focused on potential reforms in work organization to stimulate productivity growth.

These reforms are currently attempting to borrow from lessons learned in manufacturing to create worker–management–owner cooperation in organizing work to focus on the production needs of specific building processes. In attempting to create innovative cooperation in organizing work, Danish construction has relied upon its traditional strength: voluntary cooperation and regulation through collective bargaining. The advantage of this collective approach to reform is that it is possible to obtain true cooperation, from the worker up through the owner, through new organizational forms and bidding procedures. The challenge is whether the social partners and their institutions will be able to create the needed new forms of workplace organization that will stimulate productivity growth.

The challenge of raising productivity levels is significant and goes to the heart of the way of organizing the sector. It is clear that new ways of working will have to be developed if the sector's productivity is going to be increased. This means both in the forms of cooperation between the companies and in the way the production process is carried out. The division of labor between the companies with many different actors involved in the process has shown its weaknesses. Instead of optimizing the different parts of the production process, the actors often end up in discussions over organizational issues, resources, and timeframes. At the sites several experiments are being tried in order to optimize the building process. Most of them have shown good results, but those results have not yet spread throughout the sector.

At the moment, different tendencies are visible. First of all it seems very difficult to break with the well-known traditions in the sector. Although they comprehend the importance of increased productivity, individual companies have little incentive to break with the traditional way of working. To some extent, changes might have to be initiated by the public authorities. In the publicly financed building projects, a range of claims could be introduced demanding that companies implement new ways of working together.

Another tendency with consequences for productivity is the concentration process the Danish sector has gone through. The three or four market leaders in the Danish construction sector attempt to control the whole value chain. The subsidiaries of the two Swedish multinationals and the largest Danish company use the strategy of having in-house control of the whole process, from buying up land to renting out the building. It can be discussed whether this really adds to increasing productivity (e.g., some argue that the customers are left with higher prices) in spite of the way it rules out some of the traditional interface problems in the building process.

The need for structural adaptation at the construction sites is becoming more and more obvious. One obstacle to changes in the work processes is the division between different professional groups on the sites, often employed by separate subcontracting firms. This division is also visible at the level of interest representation. On the employer side, there are four main employers' organizations having relations to seven unions, all representing employees in the sector. Even though there have been attempts to simplify this picture through organizational mergers and cooperation patterns, this process could be taken further. This would assist the endeavors of creating simpler work processes through removing one reason for the clear professional divisions.

The labor market organizations have, nevertheless, been able to negotiate their way through significant regulatory changes in the sector. The international trends towards further decentralization and further flexibilization have been introduced within the regulatory regime without dramatic conflicts. The parties have cooperated with the aim of ensuring the sector's and the companies' competitiveness on the domestic and the international markets. The extensive self-regulation has shown its strength in the past decades. Now it must prove its ability to search for further innovation.

NOTES

1 In the North American context, civil engineering is often referred to as "heavy and highway construction."
2 State retirement pay is financed over taxes and given to every retired person over the age of 65 (with several arrangements for early-retirement as well). The payment is enough to live on at a basic level.
3 The industrialized building methods took shape in the 1950s and were most prevalent in the 1960s. Industrialized building led to retrenchment in materials because concrete materials replaced wood, and it led to a cutback on skilled labor because a large part of the work could be done by unskilled labor.
4 The industrialized building can be viewed as "vertical industrialization" because the building industry by using prefabricated elements carries out a larger part of the total work in the building sector (AE 1997).

REFERENCES

Arbejderbevægelsens Erhvervsråd (AE) (1997) *Report from the Industrial Council of the Trade Union Movement*, Copenhagen: AE.
Atkins International (1993) *Strategic Study on the Construction Sector*, prepared for the European Commission, DG Industry, Brussels: WS Atkins International Ltd.
Boligministeriet (1998) *The Building Policy Action Plan 1998*, Copenhagen: Boligministeriet.
Bygge-, Anlægs- og Trækartellet (BAT cartel) (1995) *Aktivitetsrapport 1995*, Copenhagen: BAT cartel.
Bygge-, Anlægs- og Trækartellet (BAT cartel) (1996) *Konjunkturrapport bygge- og anlægssektoren*, January, Copenhagen: BAT cartel.
Byggeriets Udviklingsråd (BUR) (1991) *Overvejelser om bygge- og anlægssektorens fremtidige vilkår*, Copenhagen: BUR.
Clegg, H. (1976) *Trade Unionism under Collective Bargaining: A Theory Based on Comparison of Six Countries*, Oxford: Blackwell.

Danmarks Statistik (1997a) *serie Generel erhvervsstatistik og handel 1997*, no. 6 (1 May), Copenhagen: Danmarks Statistik.

Danmarks Statistik (1997b) *Employment in the Danish Construction Sector, by Sub sector, 1966–96*, Copenhagen: Danmarks Statistik.

Danmarks Statistik (1997c) *Employment in the Danish Construction Sector, Total and by Sub-sector, 1966–96 (1995 prices)*, Copenhagen: Danmarks Statistik.

Due, J., Madsen, J. S. and Strøby Jensen, C. (1993) *Den danske model*, Copenhagen: Danmarks Jurist- og Økonomforbund.

Entreprenørforeningen (1996) *Entreprenørforeningen årsberetning 1996*, Copenhagen: Entreprenørforeningen.

Entreprenørforeningen (1997) *Entreprenørforeningens konjunkturanalyse*, July, Copenhagen: Entreprenørforeningen.

European Federation of Building and Wood Workers (EFBWW) (1995) *The Strategic Conduct of Multinational Companies – A Discussion Paper for European Works Councils in the Building and Woodworking Sector*, EFBWW Multiproject Booklet no. 2, Brussels: EFBWW.

Ferner, A. and Hyman, R. (1992) *Industrial Relations in the New Europe*, Oxford: Blackwell.

Lubanski, N. (1999) *The Impact of Europeanisation on the Building and Construction Industry in Germany, Sweden and Denmark – a comparative analysis*, Mering: Rainer Hampp Verlag.

Strøby Jensen, C. (1995) *Arbejdsmarked og europæisk integration: En sociologisk analyse af den europæiske models etablering i EU*, Copenhagen: Danmarks Jurist- og Økonomforbund.

5 Canada

Labor market regulation and labor relations in the construction industry: the special case of Quebec within the Canadian context

Jean Charest

INTRODUCTION

A common thread running through the analysis of economic issues and labor relations for more than 10 years in all industrialized countries has undoubtedly been the issue of competitiveness in a global economy vis-à-vis the role of the State. This issue has fueled challenges to the interventionist state, to highly regulated and centralized systems, to obstacles to flexibility, and so on. The question that arises is whether a strongly national – even regional – construction industry can avoid these strong forces. In fact, the Canadian experience demonstrates that the systems for regulating the construction industry in several Canadian provinces have begun to undergo a process of adjustment – even deregulation – to new market conditions. However, the case of Quebec is an interesting exception to this trend, particularly because it has followed a highly regulated model despite the difficulties experienced by the industry, which was hard hit by the two deep recessions of 1982–3 and 1991–2.

This chapter presents an analysis of the Quebec case, first situating it within the Canadian economic and institutional context and then describing the industry and its labor force, the regulation of the construction labor market, and, finally, certain pressures on the Quebec construction industry and its mode of regulation. In general, this analysis shows that the construction industry has been experimenting with union–employer parity-based structures for several decades. Moreover, parity is an integral part of its formal institutions, negotiating and joint regulation practices, and capacity to change. It has often been accompanied by government policies that, to date, have preserved the main features of the system that is particular to this industry. By and large, the system of joint regulation in the Quebec construction industry has been beneficial for employers and workers as well as for the State. It has provided the industry with a stable supply of skilled labor, supported an inclusive bargaining system for all workers in terms of wage and conditions of employment, and helped to maintain relative social harmony. The construction industry still faces a number of major challenges; ultimately, it will be the capacity of the institutions and actors to face them collectively that will determine how this centralized system will evolve.

OVERVIEW OF THE CONSTRUCTION INDUSTRY IN CANADA AND THE QUEBEC ECONOMY

To understand the context of the Quebec construction industry, it is useful to start with an overview of the construction industry within the Canadian economy as a whole. This will help put into perspective the specific features of the regulation of the Quebec construction industry, which will be examined in the next section.

The construction industry in Canada

Taken as a whole, the construction industry's importance within the Canadian economy has declined over the last four decades. Indeed, the data show that, while the industry accounted for close to 10 per cent of Canada's GDP in the early 1960s, this percentage then fell quite steadily, from an average of 7.4 per cent of GDP in the 1970s, to 7.3 per cent in the 1980s, and, finally, to 5.9 per cent in the 1990s. The same is true of residential construction, which, in terms of production value, decreased from 2.7 per cent of Canada's GDP in 1961, to 2.3 per cent in 1981 and, finally, to 1.8 per cent in 2000 (Statistics Canada 2001a).

As for the employment level, in 1961 it reached 9.3 per cent of the Canadian labor force, and then declined significantly, to reach around 6 per cent of the Canadian labor force at the end of the 1990s (Statistics Canada 2001b). Data on the sector's annual output and the number of hours worked show that labor productivity fell between the end of the 1970s and the end of the 1990s. According to Sharpe (2001), the labor-intensive nature of the industry and the low technical progress are pertinent explanations for that poor labor productivity.

The breakdown of data on construction starting in 2000 shows that the industry is highly concentrated in only four provinces. Thus, Ontario, Quebec, Alberta and British Columbia account for 90 per cent of the country's construction activity (Statistics Canada 2001c). A similar profile emerges for the distribution of the labor force. The industry's Canadian labor market is, in fact, concentrated in those four provinces, which together provide 87 per cent of jobs (Statistics Canada 2001b).

It should also be mentioned that there are significant wage gaps in the construction industry between certain provinces even when only unionized workers in the industry are considered. For example, for 1997, the wages of unionized carpentry workers (the largest trade in terms of the number of workers) varied from C\$20.98 in Newfoundland to C\$27.45 in Quebec, an average of approximately C\$32.00 in Ontario, an average of approximately C\$27.00 in the

Prairie Provinces, and, finally, to nearly C$33.00 in British Columbia (Statistics Canada 1998).

The Quebec economy and its construction industry

In terms of GDP, Quebec ranks nineteenth among countries within the Organisation for Economic Co-Operation and Development (OECD). Its GDP represents slightly more than one-fifth of Canada's GDP. With exports accounting for 38 per cent of its GDP in 1999, the Quebec economy is one of the most open economies. However, it is highly integrated into the American economy, with 85 per cent of its exports destined for the United States (Ministère de l'Industrie et du Commerce 2001). Since the North American Free Trade Agreement (NAFTA) went into effect at the beginning of the 1990s, there has been an increase in the openness of the Quebec economy in terms of foreign trade, as well as an increase in its degree of integration into the US economy.

Thus, the Quebec economy is a small, open economy with a standard of living, measured by personal disposable income (PDI), that is similar to that of G-7 countries (Canada, France, Germany, Italy, Japan, the United Kingdom, and the United States), but lower than that of Canada as a whole and that of the United States. Quebec's population is only 7.4 million, compared to 30.3 million for all of Canada, 267 million for the United States, and 94 million for Mexico; the average for G-7 countries is 97 million.

Finally, the State plays a significant role in Quebec's economy. In terms of GDP and employment, only the French State is more actively involved in the economy than the State in Quebec. The State's role is also reflected in the specific features of Quebec's labor laws. Indeed, the Quebec government is generally considered more interventionist and more protective of workers' rights than the rest of Canada and the United States. Thus, the specific nature of regulation in the Quebec construction industry must be understood from this more interventionist perspective.

Table 5.1 presents a number of difficulties facing the Quebec construction industry. First, with regard to the labor force, the industry has been declining in many respects for at least a decade. Due to the short annual working period – barely 6 months of activity on average – and the low annual wage level, construction workers are practically obliged to find other employment or another source of income to meet their needs and those of their families during the remainder of the year. In this regard, since the reform of unemployment insurance (renamed "employment insurance") in Canada in 1996, the eligibility and income support criteria have severely penalized most workers (Charest and Trudeau 2000). Based on specific studies on this industry in Quebec, approximately 40 per cent of construction industry workers are not currently eligible for employment insurance because they do not work enough hours annually (Charest 2000). The amendments to that major public policy have certainly reduced the possibility for construction industry workers to make up their annual earnings through a public income support plan. Many construction workers who

Table 5.1 Principal characteristics of the Quebec construction industry

Labor force characteristics	Employer characteristics
• 93,447 workers in 1999, 26 trades (at journey and apprentice levels); significant drop in employment since 1990 (115,659). • 857 hours (21 weeks) worked annually. • 20 per cent work 12 months, 41 per cent < 6 months. • Average hourly wage C$25.88 (+ 16.6 per cent, 1990–9), but 5.5 per cent decline in average hourly wage and 18 per cent decline in average annual wage between 1990 and 1999 taking inflation into account. • Low interregional and intersectoral mobility. • Average age 45 (journey level), 32 (apprentice level); 25 per cent of workers > 50 years old. • Annual turnover rate of 15 per cent for 1990–9. • 70 per cent work for only one employer in a year.	• 18,445 employers in 1999, 5.5 per cent more than the rest of the 1990s. • 86 per cent have less than 5 employees (average is 4 employees, decline since 1990). • Only 5 employers > 200 employees. • 75 per cent are specialty contractors (increasing constantly). • Low interregional and intersectoral mobility. • Highly developed regulatory environment (from regulation of work to regulation of conditions of employment). • Highly competitive environment.

Source: Commission de la construction du Québec 2000.

Note
Author's assessment.

do not work enough during the year must therefore consider finding another job during the industry's slack times. The high labor force turnover is an indication of how difficult it is for the industry to retain its workers for want of sufficient annual income.

The low average income in the industry is in itself a challenge for employers who are competing on the labor market with other employers who offer more stable conditions on an annual basis. It will also create problems for the industry when, given that one in four workers is over 50 years old, it soon must face the challenge of replacing part of the labor force. The traditional trades already hold little attraction for young people entering the workforce. It is unlikely that they will be convinced to work in an industry in which the expectancy of average annual earnings is lower than in several subsectors of the manufacturing industry and which, moreover, is subject to considerable cyclical variations. Employers face an environment that is highly regulated in all respects, with increasing competition for a market that declined almost continuously throughout the 1990s.

Finally, during the 1990s in particular, the construction industry – like the manufacturing industry as a whole – underwent a number of changes in terms of production techniques and work organization. First, there were changes in work tools and equipment, including the introduction of electronics. Materials are also changing considerably, as customers press for materials that are safe in every respect and employers search for ever-lower production costs. As a result of these new tools and materials, work techniques and work organization have also changed. In particular, supervision of workers has decreased, and more autonomy is required of workers on the job. Production rates have accelerated due to shorter production schedules – yet another reflection of fierce competition and the search for the lowest possible costs. Finally, some regulatory conditions of the industry have changed, with respect to, among other things, growing environmental concerns about construction sites, the materials used, and disposal of refuse. All these changes require that the labor force and employers adjust – in the former case, in order to remain employed, and in the latter, in order to remain in business.

INDUSTRIAL RELATIONS SYSTEM AND LABOR MARKET REGULATION

Governmental responsibilities in Canada are shared in such a way that labor matters in most economic sectors come under the legislative and regulatory authority of provincial governments. This is reflected in institutional frameworks that differ from province to province and in often very distinct labor relations systems. The construction industry is particularly illustrative in this regard. The Quebec construction industry, for example, can be described as highly regulated and supported by both an extensive legislative framework and highly developed parity-based structures. This section presents the main aspects of the industrial relations system in Canada and then describes the labor relations system and joint management of the labor force and training in Quebec.

The industrial relations system

For 90 per cent of Canada's labor force, industrial relations are essentially the responsibility of the provincial governments. The remaining 10 per cent – for example, federal public service employees and those working in certain industries such as telecommunications, transportation and banking – come under federal jurisdiction. For the majority, therefore, it is possible to identify differences in industrial relations regulation among the ten Canadian provinces in terms of labor laws, institutions, the organization of the system's actors, and industrial relations practices. This is particularly true of the construction industry. In general, the industrial relations systems in Canada are relatively decentralized, since the labor–management relationship is established in the first instance at the level of the firm. As a consequence, the principal union activity is the organization of new local unions on a firm-by-firm basis, with the negotiation of collective agreements for each of these units. As will be seen further on, the centralized industrial relations system in the Quebec construction industry stands out in the Canadian context.

Another difference among the industrial relations systems of the provinces is that unionization rates vary markedly. More specifically, the average overall unionization rate for the Canadian construction industry was 30.8 per cent in 2000,[1] that is, very close to the unionization rate for the entire Canadian labor force (30.4 per cent).[2] However, for the construction industry, the rate was 23.3 per cent for the four Maritime Provinces, 47.1 per cent for Quebec, 30.9 per cent for Ontario, 20.7 per cent for the three Prairie Provinces, and finally, 24.2 per cent for British Columbia (Akyeampong 2000).

On the whole, these unionization rates have been declining since the late 1970s. In fact, as a result of the 1982–3 economic slowdown, fewer large-scale public works, and a trend towards economic liberalism on the part of a number of provincial governments, the early 1980s were marked by a decline in union presence in the industry as well as a reversal of the trend towards centralized (at the provincial level) labor relations in the construction industry, at least in several provinces.

In his analysis of the Canadian industry during the 1970s and 1980s, Rose (1992) found that the 1980s were characterized by rapid growth in the use of non-unionized workers and by the use of subcontracting, often to avoid unionization. Thus, from the 1980s onwards, and particularly in Western Canada, the labor relations model that had been developed during the 1970s, which was based on a degree of centralization of relations between the parties at the provincial level for the entire labor force or for certain trades, was shaken severely. Fisher (1987) estimated that employers' use of a non-unionized labor force in an effort to reduce costs was reflected in the fact that most of the industry's activities were already being carried out by non-unionized workers, i.e., 50 per cent in British Columbia, 80–90 per cent in Alberta, 70–85 per cent in Saskatchewan and 45–60 per cent in Manitoba.

In Ontario, a centralized bargaining system for each of the industry's trades, which had been in place since 1977, resulted in a significant increase in wages, to the point where, in 1986, the wages negotiated there were the highest in North America. Because the wage gaps between unionized and non-unionized workers had become so wide, as activities slowed down in the industry as a whole, there was a relative decline in the number of construction contracts awarded to contractors whose employees were unionized. This weakened the centralized model of labor relations and gradually led to wage differences, including differences among unionized workers according to region (Foote 1987).

As will be seen further on, the specificity of the Quebec construction industry as compared with these known trends in the other provinces is very much due to the fact that its labor relations system has remained highly centralized and especially, that unionization is still mandatory for most of the industry's subsectors. This has reduced the type of wage pressures experienced elsewhere in Canada, which are inherent to competition from non-unionized labor.

The labor relations system in the Quebec construction industry

The labor relations system in the Quebec construction industry is both unified and centralized, two features that have been an integral part of its history for almost seven decades. The system is governed by a law specific to the Quebec construction industry, which explains why it has characteristics that are distinct from the construction industry in the rest of Canada and from the labor relations systems of most other economic sectors in Quebec. The latter have much in common with the American system, that is, union certification by establishment, voluntary unionization, decentralized collective bargaining, and firm-level management of labor relations with a range of possible legal remedies for the parties (in particular, with regard to dispute resolution). In contrast to this type of decentralized system, the labor relations system of the Quebec construction industry generally relies on mandatory unionization of workers, compulsory membership of employers in one of the recognized employers' associations, and on centralized, province-wide collective bargaining that determines the conditions of employment for all workers. All of these elements are included in the law applicable to the industry. Even the definition of the tasks included in the 26 trades, each of which occupies a specific, exclusive space in construction work, is specified in a regulation agreed on by the employer and union parties and included in the law. We will now examine these elements of the model.[3]

For more than 30 years, that is, from 1934 to 1969, the labor relations system in the construction industry was governed by the Act Respecting Collective Agreement Decrees, a unique law in North America that was not exclusive to this industry and that drew on European law. This law was introduced before the adoption of the Labour Relations Act in Quebec, in 1944, which created a decentralized system of representation and collective bargaining.[4] The general idea of the Act Respecting Collective Agreement Decrees was to allow, through governmental decree, the main conditions of employment negotiated by some unions and employers to be extended to all workers in a given industry, and most often on the basis of a given region. Thus, when petitioned by unions and employers in an industry, the government could decide that enough workers were covered by negotiated conditions of employment in a given industry and region and decree that the main conditions (in particular, wages) become a compulsory standard for all firms in that industry and region.[5] One of the ideas underlying this law was to prevent unionized firms from having to withstand competition from non-unionized firms. In addition, the government was seeking a way to stop the deflationary spiral of wages during the Great Depression of the 1930s.

This law was particularly well suited to the construction industry, since it established guaranteed conditions of employment on a regional basis for workers who moved from one site to another and from one employer to another during a single year. For most trades, the conditions were established on a regional basis, in 15 different regions in any given period. The application and management of each decree came under the responsibility of a joint labor–management committee financed by workers and employers. This was the beginning of the tradition of joint management that still characterizes the construction industry today.

During the 1960s, the parties wanted a new labor relations system, one that would be more "modern" and better suited to the increasing mobility of many of the industry's workers, who often moved to work on large public works. It was from this perspective that legislation specific to the industry was passed at the end of 1968; the Act Respecting Labour Relations in the Construction Industry. Although this law has since been amended (for example, by integrating responsibility for vocational training in 1987), it still provides the current foundations for the labor relations system.

A specific feature of the law is that it officially recognizes the negotiating parties, for both employers and workers, and gives them the exclusive right to negotiate conditions of employment. As a result of this law, since the early 1970s, the parties have centralized negotiations of a number of conditions of employment in all trades at the provincial level, while leaving the determination of certain other conditions to local levels (for certain regions or trades). Similarly, the parties quickly created a single provincial joint committee to apply and monitor the negotiated conditions. This was the start of the new centralized model of collective bargaining establishing unique, province-wide conditions for each trade, exclusive representation of workers and employers by certain officially and legally recognized associations, and joint management of negotiated conditions of employment. In the early 1970s, the government also established a fringe benefits committee to oversee group insurance and the pension plan for the whole industry. Over the years, the government had to intervene repeatedly both to establish conditions of employment – because of the parties' inability to come to an agreement, resulting in many work stoppages – and to determine new rules regarding representation of the parties and their effective power to participate in negotiations.

Currently, the labor relations system is based on the legal recognition, for the purposes of negotiation, of four employers' associations (i.e., for the residential, industrial, civil engineering and road, and commercial and institutional subsectors) and four union associations. Thus, according to this law, which is specific to industrial relations in the construction industry, every employer must be a member of one of the recognized employers' associations, and every worker must be a member of one of the recognized representative associations of employees. This mandatory affiliation is supervised by a joint central body, the Quebec Construction Commission (*Commission de la Construction du Québec*). In order to work on a construction site, both employers and workers must be affiliated. This is enforced by Commission inspectors during site visits. The labor relations system is also based on the negotiation of clauses common to all workers in the industry and agreements specific to subsectors at centralized bargaining tables for the entire industry, on the renewal of limited-duration collective agreements every

3 years, and finally, on joint management of the collective agreements and the labor relations through the Commission.

This labor relations system has often given rise to fierce rivalry between the union organizations over the right to represent and negotiate on behalf of workers. Union pluralism is legally established in the industry; construction workers may individually choose the union to which they would like to belong. Four separate unions offer relatively similar services and maintain generally cooperative relations with each other, since they are stakeholders in the same industrial relations system and the same industry regulation mechanisms. Thus, workers who are in the same trade or who work for the same employer may be members of different unions. Every 3 years, a union affiliation vote is held, giving workers who wish to change their union association the opportunity to do so. During a period determined by law, the unions may solicit construction workers, inviting them to join their ranks. Then, during the course of a few days, a secret ballot is held under the supervision of the Commission, and those workers who wish to change union affiliation exercise their right to vote, while the others automatically retain their original affiliation. This is an important issue, since the law stipulates that an association or group of associations that represents more than 50 per cent of workers in a sector may lead the negotiations with the employers' associations. As a result of this centralized labor relations system, wage rates (among other conditions of employment) are more or less identical for the same trade throughout the province, and negotiated wage increases more or less apply to the entire labor force. The words "more or less" are used here because, since the mid-1990s, industry subsectors may negotiate specific wage conditions. In reality, however, with the exception of the residential subsector, negotiations continue to be centralized for most subsectors.

Joint management of the labor force

From the time that a legal framework was first set up in the 1930s, the construction industry has continuously experimented with employer and union parity-based structures. This approach involved employers and unions in the monitoring of the conditions negotiated for the industry or a given region, even though the Quebec government had to intervene to determine these conditions. The parity-based approach was broadened in the early 1970s to include the joint management of fringe benefits. However, by the mid-1970s, the government determined that this arrangement was fraught with problems and unnecessary conflicts. It therefore created a body that would be independent of the parties, led by three members appointed by the government to manage conditions of employment for a period of 10 years. This was an important event because it constituted a break from parity-based structures, although various joint committees continued to function, but in a more advisory capacity. The new body, the Quebec Construction Office (*Office de la construction du Québec*), organized and monitored the union affiliation vote, collected employers' contributions and union dues and remitted them to the parties, oversaw construction site safety, managed labor power placement, and administered fringe benefits.

In 1987, at the end of this experiment, which had divested the unions and employers of several of their powers, the government concluded that those powers should be reintegrated into the regulation of the industry. The government itself had had repeatedly to establish conditions of employment because there was no negotiated agreement between the parties. This experience showed the government that the employer and union actors had a tendency to give up responsibility for industrial relations and collective bargaining, relying instead on regulation from the State. The government did not want this to happen and therefore returned to a joint approach that recognized the central role of the employer and union parties. Since 1987, the industry has been regulated by the Quebec Construction Commission, within which labor and employers' associations are a majority. This organization plays a major role in the industry. It applies the collective agreement and labor force hiring and mobility regulations, manages labor force qualifications, deals with questions related to vocational training, supervises union affiliation votes, and administers fringe benefits and any additional funds entrusted to it by the parties.

The importance of employers and unions in this parity-based organization is enhanced by the fact that they fund the Commission through contributions equivalent to 1.5 per cent of the wage bill, reserved solely for the Commission's overall operation. This provides the Commission with an annual operating budget of some C$50 million and a workforce of 600 employees. In addition, it is responsible for administering a C$7 billion retirement account that provides the funds for workers' pensions. These funds come from a special contribution by employers and workers that is remitted to the Commission periodically in order to ensure that pension payments are made to all workers in the industry based on the total number of hours worked over the course of their individual careers. The Commission also manages workers' individual vacation accounts, which are funded through another special contribution made by employers and remitted to the Commission on a regular basis. Thus, when it is time for workers to take their vacations, they receive vacation pay, directly from the Commission, equivalent to 6 per cent of hours worked plus 5 per cent for statutory holiday.

Any employer may apply to the Commission at any time to obtain the workers it needs in a given region and for the desired trades. The Commission administers a register of workers who are active in the industry and available for work. It is therefore able to provide employers with the names of available workers in the region and in the trades requested. However, most employers do not use this service but instead take care of hiring workers directly. Very often, these workers follow an employer from site to site; 70 per cent of workers work for a single employer during the course of a year (see Table 5.1).

In brief, not only is the Commission responsible for applying the law specific to its industry but it also acts as a kind of human resources management department for all workers and employers. The total contributions of employers for benefits (i.e., pension, insurance, vacation, and statutory holidays), training (i.e., through the financing of training fund), and the Commission's overall operation equal approximately 25 per cent of the average hourly wage.

The training issue

One of the main responsibilities of the Quebec Construction Commission, through a specific joint committee on training, relates to initial and further training of the industry's labor force. First, all workers who are active in the industry must hold a competency certificate issued by the Commission. This mandatory certificate ensures that only qualified workers (or apprentices) may work in the industry and that the organization of work is based on distinct, exclusive job specifications for each trade. It therefore has the power to recognize and further verify occupational skills of the labor force. Entry into the industry generally requires basic academic training, which is provided by secondary schools, and a period of apprenticeship on the job of 1–5 years, depending on the trade. Employers and unions had long demanded and, in 1987, were finally given responsibility for defining academic prerequisites for the apprenticeship system. The Commission therefore does not act only in an advisory capacity to the Department of Education (*Ministère de l'Éducation*) but also in a decision-making role as regards the content of programs – a role which is practically non-existent in any other industry in Quebec. Similarly, the Commission is responsible for defining further vocational training needs to ensure that skills are regularly updated. In brief, although the Commission itself does not provide the training, it is at the center of the regulation of skills in the industry, both in terms of entry and upgrading of the labor force.

Over the years, the Commission and the industry have adopted two powerful tools to assume the important task of upgrading labor force skills. The first tool is used to determine needs. An annual process of identifying labor force needs is jointly managed in approximately 30 subcommittees, each dealing with a trade in the industry, and in nine committees that function on regional bases. This complex process, involving approximately 300 persons in the industry, is used to determine the annual demand for further training and to plan how the training supply will be managed in several regions, in particular through public educational institutions. Courses are offered to workers who register on either a full- or part-time basis, depending on their availability and preferences. The course supply takes into consideration the industry's less active periods in order to make it easier for workers to participate.

The second tool adopted by the industry was the establishment of a training fund (the *Fonds de formation de l'industrie de la construction*) which is financed by a contribution from employers of 20 cents per hour worked by each worker. This fund generates approximately C\$18 million per year. This fund is formally under the Commission's responsibility, but a union–employer management committee decides how to use the funds collected. This fund allows employers to request training activities for their employees, and workers to register for training activities that are planned as a result of the process outlined above. Thus, the fund is reserved for those working in the industry rather than for training potential workers, a responsibility that comes under the state-financed public education system. Although the fund does not replace the wages of workers who participate in training, it does pay for the training activities themselves and reimburses workers for some of the expenses incurred (e.g., transportation and lodging), thus considerably alleviating the constraints that often prevent workers or employers from engaging in further training. According to the most recent data, for 2000–01, the system for planning training needs and the training fund is able to meet approximately 10,000 training needs (client-courses) for the approximately 76,000 construction workers eligible.[6]

No other system of this type appears to exist in the North American construction industry. The Quebec industry's training fund experiment is still in its early days; it is certainly too soon to conclude that a training culture has been established in the industry.[7] However, the involvement of management and unions in this fund and their commitment to further training suggests that such a culture will rapidly develop. The parties' commitment has been amply demonstrated through work carried out on their behalf in 2000 to help develop the fund and develop tools for its management.

The political and economic poles of the model

This level of centralization of powers within an employer and union parity-based model may seem strange in the North American environment, in which labor relations are decentralized and market rules predominate in regulating labor markets. Nevertheless, despite the demands of joint management, the partners themselves (or at least most of them) are still attached to this model. In fact, this model is part of a long historical perspective which in itself influences the way in which it is regarded by the actors (a sort of institutional "lock-in," as the institutionalists would say). However, the model both gives a considerable number of powers to the main actors themselves and allows for the collective management of the individual needs of employers and workers in a market largely made up of very small employers (with, on average, four employees per firm) and of workers who have neither job security nor legal ties that would ensure a long-term employment relationship with a particular employer. In our opinion, these two poles – one political and the other economic – explain the strength of this model.

First, the "political" pole guarantees power for the main actors. In such a system, which has a legal framework but is essentially self-regulated by employers and workers through their respective associations, the main industry and labor market actors are involved in decision-making and thus can determine certain industry conditions. This is true both of the collective bargaining process, which establishes conditions of employment, and of joint management via the Commission of everything related to the industry, from the definition of trades to the management of fringe benefits, etc. Despite the sometimes-cumbersome nature of this type of model, the fact nevertheless remains that the industry's actors are responsible for regulating the industry.

Second, the "economic" pole guarantees the distribution of beneficial services to the industry's members. In economic terms, the Commission may be considered as a collectively financed "public good," financed by employers and workers, which offers services that many employers or workers would not be able to obtain on their own. In the context of a highly fragmented industry made up of small employers and characterized by a degree of job insecurity, such an organized and centralized model provides economic and organizational conditions that help stabilize the industry and all of its members, both employers and workers. Indeed, common wage levels within the same trade, established occupational standards and labor force competency certificates, job descriptions for the trades, access to labor force placement services, the supervised application by the Commission and its inspectors of negotiated employment conditions, the management of employee benefits, and training by joint, collectively financed bodies are all advantages both for employers and workers.

In other words, the model allows the industry collectively to organize its labor market, within which the actors have considerable leeway in the choice of either their workforce, in the case of employers, or their employer, in the case of workers. The overall conditions make it possible to avoid wage competition, which could have a negative impact on the workers' standard of living and, undoubtedly, on employers' access to a sufficient number of qualified workers. Admittedly, there are constraints in such a highly regulated model – besides levies on the wage bill, there are some obligations, like the need for continuous renegotiation, which are inherent in joint, consensual management – but as one of the founders of American institutionalism, John Rogers Commons, suggested, institutions are "a collective action in restraint, liberation, and expansion of individual action" (Commons 1934: 73).

One of the strengths of this model is undoubtedly its universal character in that it applies unconditionally to all members of the industry, both workers and employers. As in the management of any public good, the effectiveness of the model is due to the fact that there are, for all practical purposes, no free-riders operating outside the industry's established norms and thereby, threatening its existence. The experience of other Canadian provinces in this regard illustrates that, as soon as employers are able to hire non-unionized workers (at lower wages than unionized workers), the centralized systems of industry regulation weaken, thus creating downward pressure on wages and unionization rates. In the following section, we will see that such threats also exist in Quebec.

PRESSURES ON THE INDUSTRY AND ON LABOR MARKET REGULATION

The description of the Canadian industry and its modes of regulation revealed sources of tensions, which we will now examine more closely. These tensions increase with regard to the highly regulated system of the Quebec industry. In addition, certain government actions would suggest that there will be modifications in the coming years to the general conditions that regulate the Quebec construction industry. We will consider three issues that are linked to the economic and political poles of this model: (1) cost control and its particular effects on wages; (2) labor force adjustment; and (3) pressures on the labor relations system. It should be remembered that the construction industry has evolved in a context of relative decline in the Canadian economy – in certain cases or periods, an absolute decline – brought about by severe recessions. In brief, the more internal nature of the market has not exempted the construction industry from the pressures experienced by industries exposed to external competition for the last 10–20 years.

The competitive environment that has resulted from a shrinking market, the continuous battle against inflation by monetary authorities, and a large number of contractors has put considerable pressures, particularly since the beginning of the 1990s, on the control of production costs and prices in the industry. A number of indicators and direct effects of these pressures can be seen. First, as was revealed in the section on the Quebec economy and its construction industry, on average, the growth in hourly wages in the industry was lower than inflation during the 1990s. Thus, despite a union presence in Quebec that has saved construction workers from having to compete directly with non-unionized workers, as is the case in several other Canadian provinces, the purchasing power of the industry's workers has dropped.

Another indicator of the downward pressure on costs (and, in particular, on wage costs) is the repeated attacks by certain employers within the association of residential builders against uniform wage rates established by collective agreements. A first attempt led the government to adopt, in 1993, a legislation that deregulated the residential sector of buildings with 8 or fewer dwelling units (which, in fact, represents 85 per cent of this sector), even though it stipulated that some negotiated conditions would be maintained for a period of 1 year.[8] This important gain for residential builders was, however, short-lived since, with a change of government in 1994, the law was abrogated and the residential sector was returned to the regulated system of collective bargaining. Moreover, the reintegration of this subsector gave representative associations the possibility to negotiate conditions specific to their subsector. In this way, the government decentralized the collective bargaining system that had been established for many years, which is what the residential subsector was trying to exploit. Actually, during the 1995 collective bargaining, the employers' association of the residential subsector decreed a lock-out, thereby eliminating the existing conditions of employment. According to data from the Quebec Construction Commission, the effect of this attack by employers was to reduce wages by approximately 10 per cent, although 55 per cent of residential construction workers continued to receive the rates established under the old collective agreement. Arbitration was required to settle the dispute in 1999; this restored the wage levels under the old collective agreement with wage increases that, nevertheless, resulted in wages that were 3–5 per cent lower than those in the other subsectors (Commission de la construction du Québec 2000).

A final phenomenon that illustrates the downward pressures on costs in the industry (especially in residential construction) is clandestine work, which increased markedly during the 1990s. Some employers reduced construction costs by paying workers less than the wages negotiated in the industry and not declaring these workers officially. While it would be contentious to try to establish the scope, in economic terms, of a phenomenon that is not subject to accounting, it can be said that it hurt legitimate workers and employers in the industry as well as the government (through lost tax revenues). The damage was enough that, in the 1990s, the government made it a priority to wage a battle against clandestine work. It allocated specific resources to the Quebec Construction Commission to strengthen its methods for monitoring and auditing industry projects. This is a good example of how free-riders threaten the survival of a model built on the logic of a public good. This explains the willingness of the unions, and the employers in particular, to expend both energy and resources in the fight against clandestine work.

A second major "economic" issue facing the industry is labor force adjustment (i.e., training and retraining), along with recruitment.[9] First, the industry's labor force is aging, one in four workers being over the age of 50. This will require more hiring over the next decade. Moreover, the turnover rate is high, 15 per cent of the labor force having been renewed annually from the 1990s onwards. Finally, in recent years the industry has had a difficult time recruiting workers with the basic academic qualifications established by the industry, which has resulted in the recruitment of workers who do not have the specific training needed by the industry. For example, for the first three quarters of 2000, of the 5000 new apprentices hired in the industry, only 2000 had diplomas. It will therefore be up to the industry (with its own training resources) to make the additional effort to upgrade the entry skills of the other 3000 new workers so that they will be able to work more effectively and improve their skills in the coming years. A number of trades are, in fact, finding it hard to recruit workers, making it even more likely that they will use workers who do not have the basic formal skills. In addition to this, the changes in materials, work methods, equipment, and regulations are also factors that make it necessary for the industry's labor force to adjust continuously.

The results of our survey, conducted in 2000, of industry leaders and a statistically representative sample of workers indicated that employer and union leaders are very aware of the importance of labor force adjustment in the industry (Charest 2000). Employers consider this a major condition for maintaining the competitiveness of firms, while the unions see it as an essential condition for maintaining the employability of workers. Nevertheless, the two parties agree that efforts at labor force adjustment are still inadequate and the challenge for the next decade will be to develop a "training culture" in the industry. Currently, approximately 3 per cent of workers participate annually in further training; the parties would like to see this proportion reach 10 per cent in the coming years. Our survey of workers revealed that the majority of those who have participated in further training in recent years recognize that it has had a positive impact on their skills, employability, and hours actually worked. The vast majority of workers surveyed attributed great importance to further training, a fact that is in itself good news for the industry. In this regard, the industry's training fund is a powerful lever for creating conditions likely to motivate workers to participate wholeheartedly. Developments over the next decade will be crucial in this regard.

To sum up, with regard to the labor force, the main challenges facing the industry are: replacement of the industry's aging labor force in the near future, recruitment of workers with the basic entry skills required in the industry, labor force stability in the industry, and finally, further training of the existing labor force, a proportion of which does not have formal academic qualifications. Moreover, the construction industry is competing for young recruits with other industries that often offer better employment conditions, particularly more stable employment on an annual basis. The construction industry therefore faces a difficult task given the data on its annual activity and on wages, and considering the pressures that may be put on wage costs in the near future.

A third and final issue facing the industry is the fact that it has a very particular labor relations system as compared with other Canadian provinces and other economic sectors. As was seen earlier, from the 1980s onwards, several provincial labor relations systems were shaken by the slowdown of activities in the industry and the increasing use of non-unionized workers. This has led to greater decentralization of labor relations, a significant decrease in union presence, and wider wage differentials. The Quebec construction industry functions in a similar economic environment but has generally maintained its centralized labor relations system, despite a few openings towards decentralization. Thus, while in the 1980s the industry elsewhere experienced deregulation, if not weaker control of the market by the regulated party, in Quebec the system ranged from less government control to greater responsibilities for employers and unions. In the late 1980s, the parties were made responsible for the joint management of the industry through the creation of the Quebec Construction Commission. The centralized system of collective bargaining has endured, often by asking the government to intervene in settling disputes, despite the fact that the industry has been in difficulty and that labor relations systems have changed in other Canadian provinces. In the mid-1990s, the government introduced decentralization by allowing separate negotiations for the four industry subsectors. This resulted in some wage differentials and especially reinforced the pro-autonomy attitude among employers in the residential subsector, who are still in favor of deregulation.

CONCLUSION

In spite of some decentralization of the system during the 1990s, the unions see the maintenance of mandatory unionization as a major gain. Moreover, the importance of unionization in maintaining Quebec's system is demonstrated by the experiences in other Canadian provinces, which have undergone the rapid penetration of non-unionized workers into the labor force. On the other hand, Quebec unions have nevertheless had to absorb a number of effects of a poor market for the industry, particularly through wage adjustments during the 1990s and the wage flexibility introduced by subsector. From the point of view of employers, decentralization of negotiations and the achievement of negotiated regulations for most subsectors, even in a difficult economic context, constitute a gain in terms of the system's ability to adjust to a new economic situation. The new collective resources introduced in the area of training also demonstrate that the parity-based, jointly regulated system is still solid.

Nevertheless, it is evident that the system is being closely observed by the government and certain employers. First, the recent attempt to deregulate the residential construction sector, which could resurface in the legislative environment in the coming years, should not be forgotten. At the Quebec socio-economic summit in 1996, the current government made a commitment to bring together the leaders of a number of economic sectors in order to start decreasing the regulatory burden throughout the Quebec economy. A result of this process to date has been the revision of the Act respecting Collective Agreement Decrees, which has existed since the 1930s and has regulated the construction industry for a very long time. In the wake of this process, several decrees establishing the conditions of employment of workers in various industries (including the large garment industry) were simply abrogated. The trend towards deregulation of the economy in the neighboring province of Ontario, Quebec's main economic partner, could put considerable pressure on the Quebec economy and its modes of regulation in the coming years. In the construction industry more specifically, pressures from employers in favor of deregulation remain great in the residential subsector.

Although Quebec's labor relations model has endured, there are definitely signs that the environment could change in the near future. However, the presence of important institutions and historically established practices between the parties act as a shield against further deregulation of the system. In fact, until now, the labor relations system of the Quebec construction industry has benefited the actors and demonstrated a certain capacity to adjust to market conditions. No doubt this is the best guarantee of its survival.

NOTES

1 The construction industry considered here for the purpose of estimating the unionization rate is defined broadly and therefore includes workers other than those directly involved in work on construction sites. It includes workers in the construction-related services subsector, engineering works, and industry promotion. This may underestimate the actual unionization rate for workers in construction as such, which is the case, for example, in Quebec where these workers are for all intents and purposes completely unionized (with the exception of home renovation – see note 8) while the rate presented here shows an overall union presence of 47 per cent.

2 The average rate of unionization in Canada disguises large differences among the provinces, for example, a rate of 21.1 per cent for Alberta as compared to 39.2 per cent for
Newfoundland and, between these two extremes, 36.1 per cent for Quebec, 35 per cent
for British Columbia, and 27.3 per cent for Ontario. It should also be pointed out that,
according to other sources and methods of estimating unionization rates in Canada
used until the mid-1990s, the unionization rates are 4–5 percentage points higher.
The data used in Quebec also result in rates that are higher than the ones cited above:
34 per cent and 40 per cent rather than 30.4 per cent and 36.1 per cent for Canada and
Quebec respectively.

3 For a detailed analysis of this system and its evolution, see Sexton 1987; Commission
de la construction du Québec 1990; Hébert 1992; Morin and Brière 1998.

4 Prior to the existence of this industrial relations system, there was already a universal
law, one that still exists today, which set a minimum wage and minimum conditions of
employment applicable to all industries and workers, unionized or not.

5 This law still exists in Quebec but has recently been reviewed by the government,
which abrogated several decrees. In a detailed analysis of this review–abrogation
process, we established that there is a direct link between this process and the context
of trade globalization. See Vallée and Charest 2001.

6 Some workers may register in more than one training activity during the year. The rate
of participation has been on the order of 3 per cent annually.

7 As a result of a legal battle waged by certain employers who were opposed to compulsory
contributions to this training fund, its activities got under way only in 1999.

8 Through legislation adopted in 1988, home renovation had already been withdrawn
from the law governing conditions of employment for the whole industry. Before that,
it had been covered by the law and the same conditions as the entire construction
industry.

9 This discussion draws on interviews conducted in the summer of 2000 with employer
and union representatives in the industry as well as on a survey on training issues,
conducted with 1300 construction industry workers in collaboration with the Quebec
Construction Commission and the Fonds de formation de l'industrie de la construction.
See Charest 2000.

REFERENCES

Akyeampong, E. (2000) "Unionization – an update," *Statistics Canada*, Catalogue
75-001-XPF.

Charest, J. (2000) *Rapport de recherche sur les incitatifs au perfectionnement et au recyclage
dans l'industrie de la construction*, Montreal: Fonds de formation de l'industrie de la
construction.

Charest, J. and Trudeau, G. (2000) "De l'assurance-chômage à l'assurance-emploi: un
baromètre de l'évolution des politiques sociales canadiennes?" *Bulletin de droit comparé
du travail et de la sécurité sociale*, Centre de droit comparé du travail et de la sécurité
sociale (COMPTRASEC), Université Montesquieu-Bordeaux IV, 66–91.

Commission de la construction du Québec (1990) *Historique des relations du travail
dans l'industrie de la construction*, Montreal: Commission de la construction du
Québec.

Commission de la construction du Québec (2000) *Analyse de l'industrie de la construction
au Québec – 1999*, Montreal: Commission de la construction du Québec.

Commons, J. R. (1934) *Institutional Economics: Its Place in Political Economy*, New York:
MacMillan Company.

Fisher, E. G. (1987) "Merit shop construction in Western Canada," in H. Jain (ed.), *Proceedings, 24th Annual Meeting of the Canadian Industrial Relations Association,* June 4–6, 1987, McMaster University, Hamilton (Ontario): 117–33.

Foote, B. M. (1987) "Bill 22 – Ten Years Later: An assessment of province-wide bargaining in the Ontario Construction Industry," in H. Jain (ed.), *Proceedings, 24th Annual Meeting of the Canadian Industrial Relations Association,* June 4–6, 1987, McMaster University, Hamilton (Ontario): 135–54.

Hébert, G. (1992) *Traité de négociation collective,* Boucherville (Quebec): G. Morin.

Ministère de l'Industrie et du Commerce (2001) "Québec's Economy is an overview of Québec Today," presentation, April 2001. Online, available at http://www.mic.gouv. qc.ca/economie/Calepin-tab-01_en.html (1 October 2001).

Morin, F. and Brière, J.-Y. (1998) *Le droit de l'emploi au Québec,* Montreal: Wilson et Lafleur.

Rose, J. (1992) "Industrial relations in the construction industry in the 1980s," in R. Chaykowski, and A. Verma (eds), *Industrial Relations in Canadian Industry,* Toronto: Dryden.

Sexton, J. (1987) "The Quebec construction industry after Bill 119," in H. Jain (ed.), *Proceedings, 24th Annual Meeting of the Canadian Industrial Relations Association,* June 4–6, 1987, McMaster University, Hamilton (Ontario): 109–16.

Sharpe, A. (2001) "Productivity trends in the construction sector in Canada: A case of lagging technical progress," *International Productivity Monitor,* Ottawa: Centre for the Study of Living Standards, 3 (Fall): 52–68.

Statistics Canada (1998) *Construction Price Statistics,* Catalogue 62–007-XPB.

Statistics Canada (2001a) *CANSIM II,* Table 379–0004.

Statistics Canada (2001b) *CANSIM II,* Table 279.

Statistics Canada (2001c) *CANSIM II,* Table 176–0056.

Vallée, G. and Charest, J. (2001) "Globalisation and the transformation of State Regulation of Labour: The case of recent amendments to the Quebec Collective Agreement Decrees Act," *International Journal of Comparative Labour Law and Industrial Relations,* Kluwer Law International, 17 (1): 79–91.

6 Australia

The Australian construction industry: union control in a disorganized industry

Elsa Underhill

INTRODUCTION

The Australian labor market has undergone significant change in the past decade. From 1983 to 1991, wages and employment conditions were determined primarily by centralized national bargaining. Agreements known as "The Accords" were negotiated periodically among the federal government, the Australian Council of Trade Unions (ACTU), and to a lesser extent, national employer organizations. These determined changes in wages and other employment conditions as well as industry policy and social policy. Changes to wages and employment conditions were given effect through legally binding awards made by Australia's industrial tribunals, which are armed with powers of compulsory conciliation and arbitration. Since 1991, this system has been replaced by decentralized enterprise bargaining as the principal means of determining wages and employment conditions. Industrial tribunals now determine wages and a narrower range of employment conditions only for lower paid workers (approximately 24 per cent of the workforce) (ABS 2000a). This decentralized system is referred to as deregulation of the labor market, and is associated with a diminished role for industrial tribunals and unions whose membership has been falling. Labor market flexibility through temporary and part-time employment has grown rapidly. Flexible working hours have become much more widespread; earnings, more dispersed.

This chapter describes an exception to the dominant national patterns of employment regulation over the past two decades. The Australian building and construction industry was never firmly regulated by centralized policies in the 1980s. Nor has it been swept along with the contemporary movement towards enterprise bargaining. A traditional regulatory approach exists in the industry, based upon strong unions and negotiated industry wages and conditions. The construction industry was never easily controlled by the centralized Accord agreements and has proven equally resilient against national enterprise bargaining policies. The chapter will show there have been significant changes in the labor market, but these owe more to idiosyncratic developments in a highly competitive industry structure than to national labor market regulation.

Underpinning the regulatory framework of the building and construction industry is the Australian political system. Australia is a federation of six states and two self-governing territories. The Australian Constitution determines the legislative powers of the federal government, with residual powers vested in the states. Of relevance to this chapter are the limited powers of the federal government to regulate the construction industry product market, the training of apprentices, and until recently, minimum wages. Regulation of the product market, including construction standards and entry into the market, falls within the jurisdiction of the state governments. The states also have primary responsibility for apprentice training, the major form of training in the industry. In both areas, the federal government's role is limited to encouraging consistency among states through consultative bodies, and through conditional funding to state governments. Competition among state governments discourages them from imposing regulations that would introduce higher costs in one state relative to others. For this reason, state governments have taken a hands-off approach to product market regulations, the effect of which is to lower barriers to entry in the construction industry. In the absence of strong national standards, inconsistencies have grown up among the states on product quality, apprenticeship training, and other issues. Since the early 1990s, however, a higher level of cooperation among states and territories has resulted in a stronger national standard for apprenticeship training, and increased uniformity in product quality regulation. Nevertheless, the industry remains substantially unregulated, with low barriers to entry and intense competition.

The federal government's power to regulate the labor market has traditionally been limited to establishing an independent industrial tribunal, the Australian Industrial Relations Commission (AIRC), for settling industrial disputes over wages and employment conditions. Until the last decade, the AIRC played a dominant role in determining minimum wages for most Australian workers. State-based industrial tribunals co-exist with the AIRC and tend to adopt similar principles with respect to the determination of wages, thereby supporting the federal tribunal. In the construction industry, state tribunals were the major forum for wage determination and dispute settlement until the mid-1970s. This resulted in industrial relations practices, union traditions, and wage outcomes idiosyncratic to each state. The construction industry labor market is now regulated primarily through the federal tribunal, although state traditions remain strong.

This chapter is divided into five further sections. The next section discusses key characteristics of the building and construction industry, including the product market and changes in industry structure. The third section describes the changing nature of employment and skill composition. In the fourth section, the nature of employment regulation and wage determination is discussed. This includes an account of union and employer representation, the methods of wage determination, and changing wage levels under a decentralized system. In the fifth section, the nature of training is discussed, with particular attention to declining training levels associated with the changing industry structure. The final section draws together the key findings from the earlier sections in a discussion of the contribution of deregulatory policies to change in the building and construction industry.

PRODUCT MARKET AND INDUSTRY STRUCTURE

The building and construction industry has accounted on average for 6–7 per cent of GDP over the last 30 years, and in 1998/99 was the fifth largest industry as a share of GDP. In terms of value added, it accounted for close to A$50 billion of economic activity in 1999. It has a high employment multiplier effect of about four (Thompson and Tracy 1993). Expenditure on building and construction is a key element in investment, so that efficiencies achieved in this industry have an impact on productive investment across other industries.

The Australian construction industry is composed of three subsectors: the housing sector, the commercial (non-residential) sector, and the engineering sector. The largest of these is the housing sector, dominated by new housing. The commercial sector includes construction of offices, shops, other business premises, and entertainment and recreation facilities, while engineering primarily involves road building and other heavy industry infrastructure such as oil, gas, and mineral processing facilities.[1] Builders tend to specialize in either housing or commercial and engineering construction, although this distinction has become blurred in recent years with commercial sector companies building high-rise inner-city apartment blocks, and expanding further into engineering construction. Buyers in the commercial sector are usually investors such as insurance companies or governments (who acquire more than 50 per cent of output). Buyers in the housing sector are usually consumers, with national home ownership rates of around 70 per cent (ABS 1999).

The industry as a whole suffers from highly volatile demand. The demand for building construction fluctuates strongly and quickly in response to changes in the level of macroeconomic activity and is considered a leading economic indicator foreshadowing wider trends. Figure 6.1 shows fluctuations in the real value of construction work, by sector, from 1985 to 1999. The engineering sector, influenced by longer-term investment projects and government infrastructure, is the least volatile. The housing sector, subject to consumer confidence and interest rate variations, and shorter project delivery times, shows the most volatility, with output varying by as much as 25 per cent over 2-year periods. The commercial sector is subject to longer troughs and booms that are strongly influenced by overall business confidence and economic cycles. Sharp and unanticipated downturns in activity have an impact, particularly heavily, on the profitability of small firms, undermining their economic viability. Larger firms are less susceptible to fluctuations in activity, being cushioned to an extent by the practice of outsourcing production activity and risk to smaller firms.

The product market for construction is highly localized, with the final output being produced "on location." In the short term, the number of firms, the available labor, and the demand for the product are confined within a given

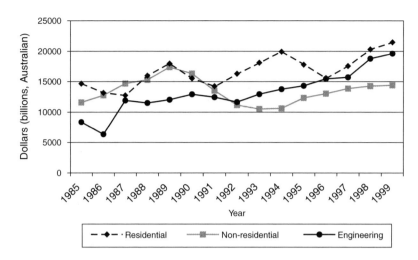

Figure 6.1 Building and construction activity by sector, 1985–99 (in 1998/99 dollars).

Sources: ABS 2000b,c.

local market. The housing sector is the most localized of the three sectors. Small firms operate within a small region, while the larger firms tend to operate only in one state, and often only in one region within that state. The major commercial construction firms have evolved from primarily state-based to national firms over the past 15 years. Paralleling this change has been a shift from large private family businesses to a significant proportion of the largest construction companies becoming subsidiaries of overseas-based construction companies. This has facilitated the expansion of these firms into overseas markets, particularly the Asian construction market, which contributes up to 20 per cent of operating revenue in some firms. Notwithstanding the national structure of many construction firms, in the commercial sector local traditions contribute to state-specific employment practices, wage levels, and union behavior.

The commercial construction sector is highly competitive. Construction projects are allocated primarily by tendering processes based upon detailed design documentation. Clients select a builder on price and quality considerations, although the lowest price bid tends to succeed. A 1995 survey of major construction companies in the state of Victoria found 60 per cent obtained almost three-quarters of their work through competitive bidding. An important consideration in both the decision to bid and the bid mark-up was the need for work (Chin and Mills 1995). Hence, when the volume of work is low, price competition intensifies as an increasing number of companies bid for fewer projects with lower profit margins. At times, a company's bidding price may even exclude a profit margin in order for the company to sustain sufficient work to cover fixed costs (Shash 1993). Profit levels for all firm sizes in the industry tend to be lower than the all-industry Australian average (Underhill, Worland and Fitzpatrick 1997: 51). Nevertheless, the level of return on investment, while subject to marked fluctuations, is often higher than in other industries due to the low level of capital investment (Productivity Commission 1999: 18). Once a contract is won, builders sublet the contract through further tendering processes, while retaining overall responsibility for the project. While some builders develop preferred subcontractor lists based upon past performance, cost considerations generally dominate selection throughout the subcontracting chain.

In the early 1990s fixed price contracts, whereby the builder bears the risk of cost overruns, became the norm. Prior to this, builders were able to pass on additional project costs to clients through time extension and price variation clauses in contracts (Barda 1995). Under fixed price contracts, cost increases due to factors beyond the control of the builder (such as design problems) are borne by the client, while site-specific cost increases are borne by the builder. Costs flowing from factors beyond the control of both parties are shared between the parties (Critall 1997). The implications of this risk sharing for the management of labor are important. Since the costs of all site-specific labor disputes are borne by the builder, the builder has a strong incentive to ensure disputes are minimized and subcontractors managed so as not to contribute to site-related industrial problems. When a delay is directly attributable to a specific subcontractor, builders pass on any additional costs they bear to that subcontractor. They may also remove the subcontractor from the project altogether. The passing-on of such costs can bankrupt a small subcontractor, placing pressure on them to conform to the builder's requirements.

The economic characteristics of building and construction outlined above are associated with the evolution of a particular kind of industry structure in which principal contractors reallocate most production to a number of subcontractors who have no direct relationship with the client. Subcontracting has become the dominant structural characteristic of the construction industry. It is endemic in the housing sector where almost all workers are now self-employed. It is found in the engineering sector, although anecdotal evidence suggests the larger scale of site employment, the geographical remoteness of many sites, and the longer duration of jobs tends to minimize its incidence. Most important has been its growth in the commercial construction sector.

Subcontracting was initially associated with trade specializations, such as electrical and plumbing contractors (Evatt 1975). Competitive pressures, particularly related to savings on labor and overhead costs, and scant regulation of entry into the market ensured its expansion. Cost-driven, non-specialist subcontracting appears to have first emerged during a building boom in the mid-1950s. By the early 1960s, declining standards of on-site amenities were attributed to the growing number of subcontractors, and disputes over subcontractors undercutting builders' wage levels were common. In response, collective agreements placing responsibility for overall site safety and amenities with the principal contractors emerged in the mid-1960s, and were enforced by unions closing the entire building site when a subcontractor reneged on safety or an agreement (de Vyver 1970). By the mid-1970s chain subcontracting, the repeated subletting of a contract or portion of a contract, had evolved across commercial building sites (Building Industry Investigative Committee 1983). Industry-level collective agreements developed in the late 1980s sought to cap the extent of subcontracting on major building sites through placing limits on the re-letting of contracts and allowing only the subcontracting of specialist tasks. This has lessened the re-letting of contracts for labor, but the definition of "specialist tasks" is very broad and the application of the agreement flexible. Extensive multiple subcontracting, involving all stages of production, now exists on all Australian commercial construction sites.

Barriers to entry for subcontractors are low. Establishment costs are minimal because subcontractors hire rather than purchase capital equipment. Registration requirements are also easy to satisfy. Builders are required to register in order to obtain building permits. While registration requirements vary among states and the nature of construction work, the Victorian example is broadly representative. Both commercial and housing builders can obtain registration after completing a tertiary qualification and 3 years work experience, or 10 years work experience with no trade or tertiary qualification (Building Control Commission 2000). Subcontractors are not required to register unless they operate as builders in their own right or practice a trade subject to licensing, such as electrical and gas plumbing work. Builder and trade registration requirements stem from quality and safety concerns, rather than attempts to restrict supply. Illustrating the concern with quality is the requirement that registered builders insure against structural defects. Such quality regulation does not impede the availability of subcontractors, most of whom are not builders in their own right and are not required to register.

The commercial sector now has an industry structure consisting of a small number of large building companies with few on-site employees, and a large number of very small subcontractors who employ the majority of workers. Subcontractors specialize in very narrow activities, such as rigging, form work, door-hanging and the like, so that a major construction project can involve up to 200 subcontractors (Productivity Commission 1999: 12). Subcontractors, in turn, move their workers from one construction site to another as required. For some, this may result in moving workers on a daily basis or even within a day. The role of the larger building companies, or principal contractors, is thus to manage the construction site by coordinating subcontractors, oversee site safety, and ensure collective agreements are complied with by subcontractors.

Some specialist subcontractors, such as air-conditioning installers, maintain a medium-size workforce. However most subcontractors are much smaller. In the early 1970s, subcontractors employing fewer than 10 persons were estimated to employ about 5 per cent of the total workforce, while firms with more than 150 employees employed just under half of the industry workforce (Evatt 1975). By the late 1980s, firms employing fewer than 5 persons had increased their share of industry employment to 43 per cent, and to about 70 per cent by 1996/97. Approximately 182,000 firms of this size operated in 1996/97, producing

54 per cent of industry gross product (Croce *et al.* 1999: 72). The next largest firm size grouping are those employing between 5 and 20 employees. In 1996/97 they made up less than 6 per cent of all firms but employed about 18 per cent of all employees. A corresponding decline has occurred in the number of larger firms. Seventeen firms employed more than 1000 employees in 1998, compared to 23 in 1986 (including non-construction employees for conglomerate firms). Of these, six operate predominantly in the commercial sector, with the remainder primarily in engineering construction sector (Toner, Cooper and Croce 2001: 85). Accompanying the shift to smaller firms has been a lowering of output per person employed in the smallest firms. Diseconomies of scale in smaller construction firms, work scheduling complexities associated with multiple subcontractors, as well as the likelihood of some self-employed contractors being partly unemployed, have contributed to declining per capita output in these firms (Croce *et al.* 1999; McGrath-Champ 1996).

Government inquiries into the construction industry have identified the following factors as contributing to the growth and continuance of subcontracting:

- Builders with a permanent workforce find it uneconomic to bid for more distant work.
- A permanent workforce cannot be fully employed during a downturn in building activity.
- Construction work requires different trades at different stages of the construction project, impeding continual employment.
- With only a small permanent workforce, builders can concentrate on management, require less fixed and working capital, and gain greater flexibility in the type, size and location of jobs to be undertaken (Industry Commission 1991; Productivity Commission 1999).

The shift to fixed-price commercial contracts over the past decade has further encouraged principal contractors to use subcontractors to pass on financial risk (Toner 2000a). Lower down the subcontracting chain, other factors, mainly related to cost reduction, also apply. These include:

- In the smallest firms, administrative costs can be concealed by the use of "free" household labor.
- Labor oncosts associated with employment, such as payroll tax, workers' compensation, and superannuation (retirement pension) payments, which are estimated to be 25–30 per cent of wage costs (Toner 2000a: 295), can be avoided by smaller subcontractors through paying cash-in-hand (black market wages) rather than full legal entitlements.
- Since at least the early 1990s, firms in the construction trade services sector (with the highest proportion of subcontracting, and including site preparation, structure, installation, and completion services) have reported average net profits below average weekly earnings for employees in the industry (ABS 2000d). Some subcontractors are selling their management and "hands-on" skills for a price less than the average tradesperson's wages.

- Up to 25 per cent of firms in the Construction Trade Services Sector reported a loss between 1995/96 and 1998/99, and a further 5 per cent reported only breaking even (a similar proportion of general contractors also made a loss over the same period) (ABS 2000d). Contracts are entered into at unsustainable prices.

Finally, increased competition coupled with a recession caused building industry prices to decline in real terms throughout the first half of the 1990s, compounding cost pressures in the industry (Underhill, Worland and Fitzpatrick 1997). In these conditions, multiple subcontracting arrangements provided a way for firms to reduce labor costs and win contracts (Wainwright 2001). There appears to be no viable alternative to this strategy, given the ease of entry into the industry. In particular, the labor-intensive nature of the industry limits the scope to compete by investing in technology and plant. As a result, the highly competitive nature of the industry has created a highly disorganized industry structure apparently ill-suited to regulation directed towards the protection of employee entitlements and security.

EMPLOYMENT IN BUILDING AND CONSTRUCTION

Employment instability

The construction industry is a significant employer of labor, being the 5th largest employer nationally, and accounting for just over 700,000 workers, or 7.8 per cent of the labor force in 1999/2000 (ABS 2000e). Even during downturns, it remains a major employer, rarely falling below 6.5 per cent of the national labor force. Volatility in construction activity is reflected in fluctuating employment levels, varying by as many as 80,000 workers from year to year. During boom periods, additional demand for labor is met both by new workers entering the industry, and through increased working hours. In 1998, for example, almost 37 per cent of construction workers worked an average of 10 hours paid overtime per week (compared with the all-industry average of 15 per cent of employees working 7 hours paid overtime) (ABS 2000f). Unemployment data suggests that construction workers on average have a lower level of unemployment than the Australian all-industry average (ABS 2000e). This may be due to employees becoming self-employed when employment is unavailable (Underhill, Worland and Fitzpatrick 1997). It may also be explained by the itinerant nature of the workforce. Semi-skilled workers losing employment in the construction sector are more likely to move on to another industry similarly characterized by a transitory workforce, such as meat processing (Underhill and Kelly 1993). Finally, the high earnings available in construction during boom periods attract workers who would normally be employed in other industries, and who return to those industries when demand in construction falls.

Short-duration employment is the norm in construction, with workers predominantly hired for the duration of specific projects. In the 1980s, average employment duration was estimated to be as low as 18 weeks (ACAC 1987: 4). A 1995 national survey found that workforce reductions (mainly compulsory redundancies) occurred in 32 per cent of construction industry workplaces, compared to 27 per cent of all Australian workplaces (Morehead *et al.* 1997: 419). A smaller study of almost 500 employees in Victorian-based construction firms in 1997 found that 45 per cent had their employment terminated within the previous 3 years, and 76 per cent of these workers had been employed for less than 1 year (Underhill and Worland 2000: 8–11). Short-term employment has characterized the industry since at least the 1970s (Evatt 1975), and places the industry above the national average in terms of instability (ABS 2000g; Evatt 1975).

Self-employment in construction

The previous section discussed the development of an industry structure based upon many small subcontractors. The counterpart development in the workforce is the growth in self-employment. The major change in the nature of employment in construction over the past 25 years has been the increased number of self-employed workers, many of whom are "dependent contractors" or "contract workers" (Underhill 1991; ILO 1997). In 1978, 21 per cent of workers (including employers) in the industry were self-employed. By 2001, this had increased to 29 per cent, peaking at 34.8 per cent in 1992. Figure 6.2 demonstrates the shifting mix of employees to self-employed from 1978 to 2001. These data understate the

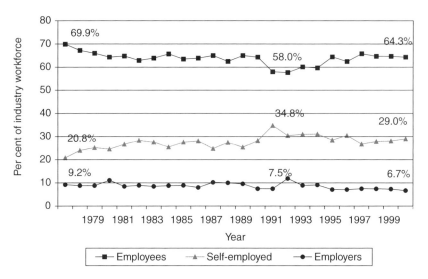

Figure 6.2 Building and construction workforce by status of employment, 1978–2001.
Source: ABS 2000e and earlier, August each year.

true number of self-employed workers. The official statistical definition of self-employed excludes those with an incorporated company. Such workers are classified as "employees" of their own company and not self-employed. Changes in taxation policy in the late 1990s led to a proliferation of self-employed workers becoming incorporated companies. They will now be officially counted as employees, even though they are employed only by themselves (Underhill, Worland and Fitzpatrick 1997).

Explanations for the growth of self-employed workers in the construction industry relate to both demand and supply factors. On the demand side, cost pressures associated with tendering provide an incentive for firms to reduce labor and administrative costs through hiring self-employed workers rather than employees. Firms hiring self-employed workers are able to avoid labor oncosts such as payroll tax, workers' compensation, and superannuation payments. They are not obliged to pay a minimum hourly rate of pay, nor are they subject to unfair dismissal legislation or redundancy payments. This provides for a greater degree of labor flexibility than is possible with an employment relationship: employers can turn the tap on and off as demand for labor fluctuates. On the supply side, significant income tax savings are available to self-employed workers but unavailable to employees. A self-employed worker earning A$50,000 can reduce income tax payments by as much as A$6000 per annum, equaling a 12.4 per cent increase in after tax income (Buchanan and Allan 2000). This provides a strong financial incentive to become self-employed. Other supply side factors include the lack of career and long-term earnings growth options for employees within the industry (associated with the predominance of small firms), as well as the desire for more "freedom and independence" among self-employed workers in the industry (Underhill, Worland and Fitzpatrick 1997). However, self-employed workers also endure less regularity of work and income (the latter compounded by the need to chase debts), are more likely to work longer hours, and rarely have income support in the event of a workplace injury. Moreover, time and cost pressures associated with payment for the job result in short-cuts, including those which jeopardize health and safety, contributing to a strong relationship between subcontracting and workplace injuries (Mayhew, Quinlan and Bennett 1996).

Many of these self-employed workers bear most of the characteristics of employees but lack their legal protections or entitlements. They have historically been opposed by construction unions, who endured membership losses when self-employment became the norm in the housing sector in the early 1970s. Unlike the United Kingdom, however, in Australia the construction unions responded to the growth in self-employment in the commercial sector by first enrolling these workers, and second, providing an industrial service to them (often through demanding back-pay on entitlements they would have received had they been employees) (Underhill 1991). In this way, the growth of self-employment has not undermined union membership to the extent that it has in the United Kingdom or Spain (see the chapters on the United Kingdom and Spain).

Table 6.1 Occupational composition of the construction industry, 1996

Occupation	Per cent of total workforce
Managers and administrators	9
Associate professionals	6
Carpenters	10
Electricians	10
Plumbers	8
Painters and decorators	6
Bricklayers	4
Other tradespersons	13
Plant operators	10
Laborers and related workers	10
Other	14

Source: ABS 1996, as cited in Productivity Commission 1999: 19.

Skill composition in the industry

Workers in the construction industry are predominantly tradespersons (51 per cent) who have completed an apprenticeship, followed by those without a post-school qualification (42 per cent). There are relatively few professionals employed compared with the national workforce profile (Construction Training Australia 1999: 12). Table 6.1 gives the occupational composition of the construction industry workforce in 1996.

Compared to 30 years ago, when 30 per cent of on-site workers were carpenters (Evatt 1975), the skill composition of the workforce appears to have declined. Technological and material input changes, such as the transfer of trusses, window frames and the like to off-site production, and the development of metals, plastic fibers, and concrete walls, have reduced the demand for carpenters. A shift away from formal apprenticeships towards on-the-job training during periods of skill shortages has also contributed to a decline in the number of formally qualified carpenters. The proportion of other trades, with the exception of a marked decline in bricklayers, has remained relatively stable over the last 30 years (Evatt 1975). A long-term decline in formal training has been evident in the building industry since the 1970s (Hutton 1970), and is widely regarded as reaching a crisis point. This is discussed further in the training section below.

EMPLOYMENT REGULATION AND WAGE DETERMINATION

Unions and employers' associations

Unionism in building and construction was traditionally craft-based, and relatively localized by state. The Victorian situation typified that in other

Australian states. Until the late 1980s several state-based unions covered brick-layers, tilers, and plasterers, while a number of national unions covered painters, plumbers, electricians, carpenters, crane drivers, and laborers. The national building unions had a history of state autonomy, operating almost as separate unions in each state. In the early 1990s, a wave of amalgamations consolidated unionism in every industry, including building. As a result, one major union was created covering all trades and laborers on building sites, with the exceptions of plumbers, electricians, and metal workers. That union, the Construction Forestry Mining Energy Union (CFMEU), also organizes independent mining, forestry, and electricity divisions, as well as building workers. Within the construction division, the states retain the autonomy they have always had.

The level of union membership remained relatively high in the construction industry until the 1980s. A decline in membership has since occurred, consistent with the general fall in union density of the Australian workforce as a whole (Peetz 1998). Current membership is about 26 per cent of the construction workforce, with considerable differences in unionization rates believed to exist among states, and outside capital cities (ABS 2000h). In the commercial and engineering sectors, compulsory unionism agreements, applying to all workers irrespective of employment status, were introduced in the mid-1970s after an extremely protracted industrial campaign (Underhill 1991). Compulsory unionism was outlawed in Australia in 1996, and the building and construction industry was targeted unsuccessfully by the federal government for "special treatment" to eradicate this practice. However, employers have been unwilling to overturn established practices, and *de facto* compulsory unionism continues to exist on major construction sites. This enables unions to exercise power in the commercial and engineering sectors, but not in housing where unionism has been extinct since the early 1970s.

The industry has been characterized by a high level of industrial militancy since the early 1960s (Moran 1987). It experiences a higher-than-average level of industrial action, illustrated for example by the year 1999, when 381 working days were lost per thousand employees in construction, compared to 87 for all industries (ABS 2000i). The CFMEU derives its bargaining power both from the nature of the industry and from strong membership loyalty. First, the substantial cost of delays to projects causes builders to be more willing to succumb to union demands than their counterparts in other industries. This is especially the case when a dispute concerning one section of a construction site may spread rapidly across a whole site. Second, the CFMEU has a potent site-level presence embedded in collective agreements. This includes compulsory site induction training for all workers; a full-time union representative on all major sites; a highly pro-active approach to health and safety; and the development of training programs for semi-skilled and unskilled workers, which is supported by paid training leave provisions in collective agreements. Third, the CFMEU acts on behalf of subcontractors as well as employees in the collection of outstanding monies, mitigating otherwise potential antagonism from subcontractors. Finally, the unstable nature of employment in the industry contributes to employees developing a stronger sense of loyalty to the union than to their current employer.

A large number of employers' associations exist in the Australian building industry. A significant proportion of these represent specialist trades, such as the Master Plumbers and Mechanical Services Association of Australia and the National Electrical Contractors Association. These trade-based employers' associations participate jointly in industry-level bargaining and actively promote training and skill development in their respective fields. They also play a critical role in providing services to specialist subcontractors. General builders and subcontractors belong to the main industry associations, the Master Builders Association (MBA) and the Housing Industry Association of Australia. The first of these mainly enrolls builders in the commercial sector of the industry, while the latter is restricted to the housing sector. The MBA is the oldest of the employers' associations, and is generally recognized as the main employer representative in the industry. It represents employers in collective negotiations and appearances before industrial tribunals and is influential in lobbying governments.

The extension of an articulated subcontracting industry structure has destabilized the coherence of employers' associations in the industry. In the late 1970s, principal contractors broke away from the MBA to form their own association to negotiate with unions, restricting the MBA's membership to smaller subcontractors (Rose 1987). That association of principal contractors later dissolved as the principals began negotiating individually rather than collectively with unions. A "Building Industry Developers" group, an ad hoc association, has since formed in one state to represent major principal contractors in industry collective negotiations (Incolink 1996). Notwithstanding these associations, principal contractors tend to break away from employer groupings when placed under industrial pressure by unions.

Wage determination in construction

Until the mid-1970s, wage rates for tradespersons in the building and construction industry were determined in awards made by state tribunals, while laborers were paid under a federal award determined by the AIRC. State tribunals were empowered to arbitrate wage rates, following submissions by unions and employers, and to apply those rates to all employers in the industry irrespective of whether their workforces were unionized. The AIRC operated on a similar basis, but without the power to issue common rule applications, so their awards applied only to named employers. Thirty-five state awards existed in 1975 (Rose 1987: 472), made for each of the many state-based craft unions and autonomous state branches of national unions. Wage increases granted in the bigger states were typically used as the basis for wage increases in the "follower" states. State awards were more attractive than federal awards, as their broader coverage enabled unions to apply minimum rates and conditions to the smaller subcontractors often excluded from federal awards. The first national award to cover tradespersons in construction was established in 1975. This national award continued to distinguish among states for some rates of pay, maintaining traditional interstate wage relativities. Laborers continued to be covered by a separate national award until their union was outlawed, in most states, in the mid-1980s. Laborers are now also included in the same national award as tradespersons (with the exception of a minority of states where the laborers' union continues to operate). Until recent times, the only industry-wide collective wage agreement covering more than one trade existed in Victoria, where an informal agreement was first reached in the late 1950s and renegotiated at irregular intervals (de Vyver 1970). In all states however, site agreements were negotiated for each major construction project. These agreements included allowances specific to the nature of the project (such as dirt and height allowances), compulsory unionism, and applied to all workers, whether they were employed by the principal or by subcontractors. From the mid-1970s, employment in the building industry was therefore regulated through a mix of agreements and federal and state awards. The awards of the tribunals set the minimum terms and conditions of employment, to which industry and project agreements were added.

This complex web of awards and agreements continued in the construction industry throughout the 1980s, when most other Australian workers were covered primarily by awards of industrial tribunals, rather than formal collective agreements. The heavier reliance on the tribunal system by other industries, coupled with policies agreed between the federal government and the peak federations of unions and employers, resulted in a highly centralized wage determination system from 1983 to 1990. This centralist policy, which provided for improvements in social benefits and taxes to offset centrally disciplined wage restraint, was implemented mainly by the AIRC. The building and construction industry continued to negotiate site agreements, and extend and improve the industry-wide agreement in the state of Victoria, throughout this period. The industry was never firmly regulated by the centralist policies dominating wage determination in other industries.

Beginning in the early 1990s, legislation and policy governing wage determination in Australia was amended to remove centralized wage determination through awards and to encourage company-level bargaining. Awards are now only a "safety net" for employees unable to negotiate collective agreements. The CFMEU has responded to enterprise bargaining by developing comprehensive enterprise agreements with individual principal contractors, and state-level "pattern agreements" to cover all subcontractors. These pattern agreements are negotiated with employers' associations and set standard conditions and pay rates across the industry. While agreements with principal contractors offer marginally higher rates of pay than the pattern agreements, the small on-site workforce attached to principal contractors means that the state pattern agreements are now the main device to fix pay.

The enforcement of collective agreements takes place through several processes. First is union enforcement, which includes the right of the union to inspect payroll documentation; second, collective agreements with principal contractors require that only subcontractors covered by pattern agreements be hired on principal contractor sites; and, third, principal contractors require subcontractors to abide by pattern agreements as part of their tendering process. Subcontractors, particularly small operators entering a construction site for only a few days, can still slip through this web of enforcement, but the penalties for getting caught are high, including prohibition from further entering the site. Finally, legal proceedings for breaches of collective agreements generally occur only as a last resort and often only with respect to underpayment of wages (the exception being breaches of a precedent-setting nature associated with broader anti-union strategies of employers). The court process is considered by unions to be both too costly and too slow for enforcing agreements.

Wage rates in construction

Wage rates in construction are influenced by the level of building industry activity, and increasingly by union bargaining power as collective bargaining displaces award rates of pay. Volatile earnings (inclusive of overtime payments) have long been a feature of the industry, enabling it to draw in new workers when demand increases. However, superimposed upon this cyclical pattern are trends favoring the construction industry as a whole, and, in particular, those workers under enterprise agreements. Since the introduction of enterprise bargaining, the construction industry has consistently negotiated agreements incorporating average annual wage increases higher than the national all-industry average (DWRSB 1998; DEWRSB and the Office of the Employment Advocate 2000).

Figure 6.3 shows the ratio of average weekly earnings of full-time adult employees in construction compared first to manufacturing (with a similar workforce composition of trades to non-trades) and second to the all-industries average. Average weekly earnings includes all allowances and payments for additional hours worked. Variations in construction industry relativities match peaks and troughs in the level of construction industry activity (see Figure 6.1). Therefore, relative pay in construction declined as the industry stalled during the early 1990s before recovering for the remainder of the 1990s. Overtime payments ensure average weekly earnings are sustained above the all-industry average, particularly during boom periods.

The growth of enterprise bargaining throughout the 1990s has seen the emergence of a union/non-union wage differential as a unionized enterprise bargaining sector has forged ahead of non-unionists reliant on arbitrated national safety net adjustments (ABS 2000a). This national pattern is beginning to emerge in the construction industry and reflects differences between the heavily unionized capital city projects in the various states, and the more weakly unionized outer-suburban and rural areas. Official statistics distinguishing union and non-union earnings have not traditionally been collected in Australia due to the dominance of the award wage system which applied minimum terms and conditions to employees irrespective of union membership. In 1999, such data

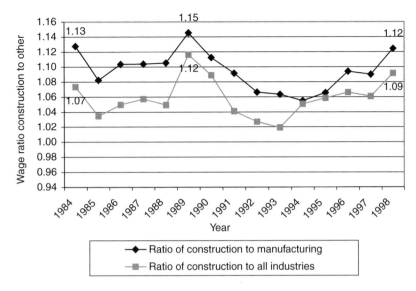

Figure 6.3 Average weekly earnings ratio, full-time adult employees, construction to manufacturing and construction to all-industries average, 1984–99.

Source: ABS 2000f, August each year.

was collected for the first time in recognition of the newfound diversity of wage determination methods. In construction, the wage differential for union members was 5.5 per cent compared to 3 per cent in all industries (ABS 2000a).

Industry-specific employment benefits

Many basic conditions of employment in construction are akin to those in other Australian industries. Employees are entitled to 4 weeks of paid annual leave, 10 days of sick leave, and similar standard hours. Unlike the United States and some European countries, pension and social fund contributions have not historically been included in collective agreements nor developed at the industry level in Australia. The construction industry differs from the national pattern in this respect. An extensive array of portable, industry-based funds have been collectively negotiated to enable employees to accrue employment benefits based upon employment in the industry rather than with a single employer. Such portable benefits are important in an industry where short-term employment is pervasive. They include long service leave, established in 1977 and supported by state legislation throughout Australia. This entitles workers to 13 weeks of paid leave after 10 years of continuous service in the industry. When an employee leaves the industry for more than 2 years, entitlement for that person ceases and the contributions made to the fund on that person's behalf are retained by the industry fund. Initially employers made regular contributions based upon their wage bill. By 1993, the Victorian fund had become fully self-funding, and employer contributions ceased. Second is superannuation, similar to a retirement pension. The construction industry was the first to develop portable superannuation, established in 1984 and now compulsory for all employers across Australia. Third is redundancy payments, whereby weekly payments made by employers into a jointly governed central fund are accessed by workers upon termination of employment. In addition to redundancy payments, the fund now supports a range of services to the industry, including training (non-trade), counseling, and industry-specific research. These schemes have been extended at a time when other industries have moved away from industry-level negotiations under deregulation. The most recent portable entitlement, introduced in 1997, is portable sick leave, which enables employees to accrue up to 100 days of sick leave entitlements (Underhill and Worland 2000). Each of these portable entitlement schemes was initially opposed by employers and introduced after protracted industrial campaigns by the building industry unions. They were introduced on a state-by-state basis, so differences among states continue in building worker entitlements, reflecting the priorities of the state branches of the national union. A number of other unions with membership mainly beyond the building industry can now gain access to the superannuation and redundancy schemes for their members who work on construction sites. Employer contributions to these schemes amounted to just under 16 per cent of average weekly total earnings in 2000.[2]

The building and construction industry as a whole shows a strong contrast between the highly unionized commercial and engineering sectors and the non-unionized housing sector. While industry disorganization caused by unregulated competition pervades both, the union succeeds in superimposing strong employment regulation on the commercial and engineering sectors. This is so despite the anarchic tendencies of the subcontracting system. High wages are protected by industry pattern bargaining, although recession conditions undercut the force of agreements; moreover, employment insecurity is mitigated by a wider range of portable benefits. However, employment regulation depends upon union strength. This in turn depends upon compulsory unionism, which remains vulnerable to political attack.

IMPERILED TRAINING

Training in the building and construction industry occurs through formal trade apprenticeships in the form of a 4-year indenture, which combines classroom attendance with supervised on-the-job training. The supply of skilled workers suffers from the highly competitive and disorganized structure of the industry, which tends to promote a short-term approach to skill development at the expense of the future well-being of the industry. Employers prefer to poach trained workers from other employers rather than pay for training themselves (Buchanan and Sullivan 1996). As a result, the number of apprentices undergoing

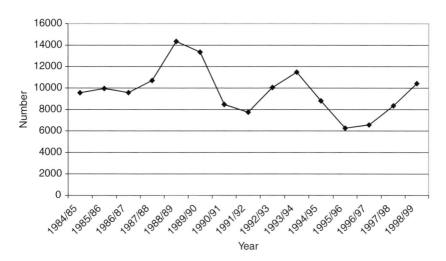

Figure 6.4 Number of apprenticeship commencements in building and related areas, 1984/85–1998/99.

Source: Toner 2000b: 43.

Note: For 1997/98 and 1998/99, apprentices are redefined to include those whose anticipated duration of training is more than 2 years.

trade training has fallen continually since the 1970s. Figure 6.4 shows the number of apprenticeship commencements in building trades for 1984/85–1998/99 for all of Australia. These include trades such as carpenters, painters, plasterers, plumbers, tilers, and refrigeration and air-conditioning mechanics. Training models in Australia have undergone considerable change in recent years. Hence, the data for 1997/98 and 1998/99 are inflated by the inclusion of apprentices with an anticipated duration of training of more than 2 years, rather than the standard 4-year trade apprenticeship.

Recognizing the shortage of skilled tradespeople, governments have provided incentives for training. Employers receive a government subsidy equivalent to approximately 4 per cent of the apprentice's wage, and apprentices commence with a relatively low wage (55 per cent of the tradesperson's basic wage), which increases over the life of the apprenticeship. However, governments have also contributed directly to the decline in apprentice numbers. Historically governments were major employers of apprentices. As they have outsourced their construction work, their share of industry employment has fallen dramatically over the last 15 years. Few apprentices are now hired in the public sector (Construction Training Australia 1999; Toner 2000b).

Fluctuations in apprenticeship numbers generally match changes in industry employment. Figure 6.4, however, shows that recent job growth has not been accompanied by a corresponding increase in apprenticeship commencements. Hence, while total employment was similar in both 1989/90 and 1995/96, apprenticeship commencements in 1995/96 were 38 per cent below those of

1989/90. It is widely believed that apprentice training in construction is entering a crisis, with current training levels falling well below the industry's long-term skill needs.

One factor blamed for the decline in apprenticeships has been the growth of subcontracting. Too few subcontractors are willing to hire apprentices. In the early 1980s, the Victorian MBA responded to this problem by establishing a "group training scheme." Under this scheme, apprentices were employed by the group training company and allocated to employers on a temporary basis according to employer demand, overcoming the requirement for an individual employer to commit to a 4-year indenture. The group training scheme expanded over time to become an organization known as the Building Industry Group training scheme (BIGS), managed on a tripartite basis by government, employers, and unions. It was responsible for the employment of almost 30 per cent of construction industry apprentices in Victoria by the late 1990s (Construction Training Australia 1999). However, the BIGS scheme collapsed in 2001, and apprentices employed through BIGS were transferred to other non-industry-specific group training schemes. The industry that pioneered group training in Australia could not sustain its own group training scheme. Meanwhile the number of apprentices in the industry continues to decline.

Another initiative to solve the apprenticeship problem is a trade union initiative on large sites where construction continues for several years. Since the early 1990s, a small number of company-level collective agreements have been negotiated, requiring the hiring of apprentices. However, their impact on apprentice levels is limited by the small number of builders willing to enter such agreements. They have not changed the way the industry as a whole approaches apprenticeship training (Buchanan and Sullivan 1996). Builders participating in these agreements continue to be at risk of being undercut by competitors who take no responsibility for hiring apprentices. To fill the gap left by employers, the main trade union in the industry has now begun offering training itself in some states, acquiring the status of an accredited provider of formal vocational training.

Two factors, cost and uncertainty, have been associated with the reluctance of subcontractors to take on apprentices. A number of studies support the importance of the second of these factors. A 1997 survey of self-employed and very small subcontractors in Victoria found strong reluctance to hire apprentices (Underhill, Worland and Fitzpatrick 1997). They considered their business environment too uncertain to risk hiring apprentices, and believed they had insufficient time to supervise and administer apprentices. Similar impediments were identified in a survey of electrical contractors, which found that economic uncertainty, rather than cost, stopped them hiring apprentices (Doughney *et al.* 2001).

Nevertheless, cost factors continue to dominate most parties' approaches to the training problem because apprenticeship training is viewed by most employers as a cost rather than an investment (i.e., in future skills). To overcome this, state governments provide subsidies to employers for apprenticeship training. However, these are now considered too low to provide an incentive (Dockery *et al.*

1997). To provide additional funding, industry training levies have been introduced. However, this approach also falls short. These schemes are confined to states where the least construction activity takes place (Construction Training Australia 1999; Toner 2000b). While a number of state-based, industry-funded inquiries into declining apprenticeship levels have recommended training levies be applied to all industry participants, few state governments appear willing to introduce such funding mechanisms (Buchanan and Sullivan 1996; Doughney *et al.* 2001). The overlap of federal and state government responsibilities for apprenticeship training compounds the difficulty in establishing an industry-wide solution to the problem.

Coinciding with the decline in traditional apprenticeship training is the emergence of a new approach to skill requirements in the industry. First, there has been an increase in the degree of task specialization among builders and subcontractors, which, associated with new materials and technologies, has caused de-skilling. Many subcontractors no longer cover the full range of tasks required for apprenticeship training. A carpenter, for example, may now work only on form work or door-hanging. Second, because the skill required for these tasks is much narrower than that taught in an apprenticeship, it can readily be learned on the job without specialist training. This approach is seen to meet industry needs in the very short term, but raises questions about the declining level of skill and quality of construction work.

The decline of apprenticeship training reflects more general weaknesses in construction industry training. Non-trade training in management and other areas of skill requirement falls below the average. In 1996, the construction industry spent just over A$100 per employee on training, compared to the all-industry average of A$185 (Underhill, Worland and Fitzpatrick 1997). In addition, literacy and numeracy skills are considerably poorer than in the rest of the workforce, although in part this is due to the higher proportion of immigrants in the workforce (Construction Training Australia 1999). The overall trends in construction industry skill formation are negative and consistent with the growing disorganization evident in industry structure.

CONCLUSION

This chapter has discussed a number of major changes in the Australian construction industry since the 1970s. The industry has maintained an important role in the economy throughout this period and appears to be relatively efficient and cost competitive (Toner *et al.* 2001). It has been characterized as highly volatile, with employment levels expanding and contracting dramatically over short time periods. Foremost among the changes in the industry has been the growth of small subcontractors. The shift to subcontracting has influenced a number of other industry trends. In particular, it has been associated with an increase in the number of self-employed workers, declining training, and the continuance of occupational health and safety risk-taking in the industry.

Significant changes in wage-fixing arrangements threaten increased wage dispersion within the industry, placing pressure on unions to adopt a more aggressive approach in order to maintain wage relativities.

What role has labor market deregulation played in these changes? With the exception of wage-fixing arrangements, the major changes identified in this study originated before deregulation of the labor market throughout Australia began in the 1990s. In particular, the development of subcontracting and self-employment reflects long-standing gaps in regulation rather than deregulation. The growing number of smaller firms has been facilitated by weak barriers to entry into the industry. Such firms do not have to be licensed to operate, nor do they need capital reserves to enter the industry. In a product market distinguished by competitive bidding and cost cutting, small builders provide an outlet for larger firms to reduce labor costs and to minimize financial risk. The shift to self-employment has also helped smaller firms to cut labor costs to survive. Both demand and supply factors have influenced this shift, facilitated by income tax legislation that benefits self-employed relative to employed workers and employment legislation that has consistently excluded the self-employed from wages regulation. In other industries, profits from cost-saving innovations can be shared among companies, employees, and consumers. By contrast, in construction, competitive bidding invariably results in cost savings flowing only to the end buyer of the product, forcing builders to continually find new ways to reduce immediate costs, irrespective of the longer-term consequences for the industry overall. Subcontracting offers an easy solution, while also raising questions about skill and construction quality.

Labor market deregulation in Australia is associated with three trends. First, it has been aimed at curbing union influence and weakening arbitrated employment standards in favor of more variable, market-driven, enterprise agreements. For the most part, the commercial and engineering sectors of construction have withstood this development, remaining strongly unionized and maintaining coordinated industry wage bargaining as well as a number of portable industry benefits. Second, deregulation of the labor market in most industries in Australia has also been tied to increasing job insecurity most often associated with increased casualization. In construction, the workforce has always been subject to a high level of employment instability, and deregulation does not appear to have influenced the magnitude of this instability. Third, deregulation has also been associated with increased wage inequality as less-unionized workers and those with little bargaining power experienced declining relative wages. Construction has shared in this trend. Wage dispersion appears to be increasing.

At a national level, the most radical labor market deregulation has occurred only since 1996. This coincides with the beginning of the present construction industry boom. It remains to be seen how outcomes will differ in less favorable economic circumstances. Coupled with an already disorganized industry structure, any decline of union bargaining power may severely test the ability of the union to continue to resist the spread of labor market deregulation into the construction industry.

Intense product and labor market competition in building and construction tends to discourage a "high road" approach in which firms seek competitive advantage through high-trust work organization and skills. The dominant trend is for firms to seek short-term profitability by undercutting to win work, and then shaving costs on skills and quality. As Kochan and Osterman (1994) argue, a high road approach requires supportive government policies. These are not found in Australia. The diffusion of responsibility between federal and state governments, coupled with the contemporary preference for deregulation, creates the wrong kind of political environment. By themselves, employer and union initiatives to support high road policies are looking increasingly fragile.

NOTES

1 In the North American context, engineering contractors are often referred to as "heavy-and-highway contractors."
2 Calculation based upon levies specified in the CFMEU Victorian Pattern Agreement 1999–2002, Clause 23.2; the Victorian Building Industry Agreement 2000–2005 (Incolink 2000: Clause 29); and Glasson (2001: 3), compared with average weekly total earnings for construction employees, August 2000 (ABS 2000a).

REFERENCES

Australian Bureau of Statistics (ABS) (1996) *Census of Population and Housing, 1996*, Cat. No. 2031.0, unpublished data, Canberra: ABS.
Australian Bureau of Statistics (ABS) (1999) *Housing Occupancy and Costs, Australia*, Cat. No. 4130.0, Canberra: ABS.
Australian Bureau of Statistics (ABS) (2000a) *Employee Earnings and Hours, Australia*, Cat. 6305.0, Canberra: ABS.
Australian Bureau of Statistics (ABS) (2000b) *Building Activity*, Cat. 8752.0, various years, Canberra: ABS.
Australian Bureau of Statistics (ABS) (2000c) *Engineering Construction Activity*, Cat. 8762.0, various years, Canberra: ABS.
Australian Bureau of Statistics (ABS) (2000d) *Summary of Industry Performance*, Cat. 8140.0.40.002, Canberra: ABS.
Australian Bureau of Statistics (ABS) (2000e) *Labour Force*, Cat. 6291.0.40.001, Canberra: ABS.
Australian Bureau of Statistics (ABS) (2000f) *Job Vacancies and Overtime*, Cat. 6354.0, Canberra: ABS.
Australian Bureau of Statistics (ABS) (2000g) *Labour Mobility*, Cat. 6209.0, Canberra: ABS.
Australian Bureau of Statistics (ABS) (2000h) *Trade Union Members, Australia*, Cat. 6325.0, Canberra: ABS.
Australian Bureau of Statistics (ABS) (2000i) *Industrial Disputes*, Cat. 6321.0, Canberra: ABS.
Australian Conciliation and Arbitration Commission (ACAC) (1987) *Application for Variation to the National Building Trades Construction Award 1975*, Case 4263, Print G8850, Melbourne: ACAC.

Barda, P. (1995) *In Principle: A Celebration of the Work of the Construction Industry Development Agency*, Canberra: Australian Government Printing Service.

Buchanan, J. and Allan, C. (2000) "The growth of contractors in the construction industry: Implications for tax revenue," *The Economic and Labour Relations Review* 11 (1): 46–75.

Buchanan, J. and Sullivan, G. (1996) *Skills Formation in the Construction Industry: Lessons from some Recent Innovations*, Australian Centre for Industrial Relations Research and Training, Working Paper 45, Sydney: University of Sydney.

Building Control Commission (2000) *Who needs to be a Registered Building Practitioner?* Melbourne: Building Control Commission.

Building Industry Investigative Committee (1983) *Preliminary Investigation into Cash-in-Hand and Pyramid Sub-contracting Practices in the Victorian Building and Construction Industry*, Report to The Premier and Cabinet, Melbourne: Government of Victoria.

Chin, T. and Mills, A. (1995) *An Analysis of Contractor's Bidding Decisions: A Study of Competitiveness of Construction Companies*, Melbourne University Construction Research Group, Department of Architecture, Building and Planning, Melbourne: University of Melbourne.

Construction Training Australia (1999) *Building and Construction Workforce 2005: Strategic Initiatives*, Melbourne: National Building and Construction Industry Training Council Limited.

Critall, J. (1997) "Industrial relations risk in the construction industry: a contractual perspective," in T. Bramble, B. Harley, R. Hall, and G. Whitehouse (eds), *Current Research in Industrial Relations: Proceedings of the 11th AIRAANZ Conference*, Brisbane: Association of Industrial Relations Academics of Australia and New Zealand, 432–38.

Croce, N., Green, R., Mills, B., and Toner, P. (1999) *Constructing the Future: A Study of Major Building Construction in Australia*, Employment Studies Centre, Newcastle: University of Newcastle.

Department of Employment, Workplace Relations and Small Business (DEWRSB) and the Office of the Employment Advocate (2000) *Agreement making in Australia under the Workplace Relations Act, 1998 and 1999*, report, Canberra: DEWRSB.

Department of Workplace Relations and Small Business (DWRSB) (1998) *Report 1997: Agreement-making under the Workplace Relations Act*, report, Canberra: DWRSB.

de Vyver, F. T. (1970) "The Melbourne Building Industry Agreement: A re-examination," *Journal of Industrial Relations* 12 (2): 166–81.

Dockery, A. M., Koshy, P., Stromback, T., and Ying, W. (1997) "The Cost of Training Apprentices in Australian Firms," *Australian Bulletin of Labour* 23 (4): 255–74.

Doughney, J., Howes, J., Worland, D., and Wragg, C. (2001) *Apprentice and Ongoing Training Needs in the Electrical and Associated Industries*, Workplace Studies Centre, Melbourne: Victoria University of Technology.

Evatt, E. (1975) *Interim Report of the Inquiry into Employment in the Building Industry*, Canberra: Department of Labour, Australian Government.

Glasson, J. (2001) "Incolink Twelve Years On," *On Site* 16: 3–4.

Hutton, J. (1970) *Building and Construction in Australia*, Melbourne: Cheshire Publishing.

Incolink (1996) *Victorian Building Industry Agreement 1996–2000*, and earlier years, Melbourne: Incolink.

Incolink (2000) *Victorian Building Industry Agreement 2000–2005*, Melbourne: Incolink.

Industry Commission (1991) *Construction Costs of Major Projects*, Canberra: Australian Government Printing Service.

International Labour Office (ILO) (1997) *Contract Labour: Report VI (1)*, Report from the International Labour Conference, 85th Session, Geneva: ILO.

Kochan, T. A. and Osterman, P. (1994) *The Mutual Gains Enterprise*, Boston: Harvard Business School Press.

Mayhew, C., Quinlan, M., and Bennett, L. (1996) *The Effects of Subcontracting/Outsourcing on Occupational Health and Safety*, Industrial Relations Research Centre, Sydney: University of New South Wales.

McGrath-Champ, S. (1996) *Employee Relations in the Construction Industry*, Australian Centre for Industrial Relations Research and Training, Working Paper 44, Sydney: University of Sydney.

Moran, A. (1987) "Hard Labor for the BLF," *Arena* 81: 21–4.

Morehead, A., Steele, M., Alexander, M., Stephen, K., and Duffin, L. (1997) *Changes at Work: The 1995 Australian Workplace Industrial Relations Survey*, Melbourne: Longman.

Peetz, D. (1998) *Unions in a Contrary World: The Future of the Australian Trade Union Movement*, Cambridge: Cambridge University Press.

Productivity Commission (1999) *Work Arrangements on Large Capital City Building Projects*, Canberra: AusInfo.

Rose, J. B. (1987) "Multi-employer cohesion in Australian construction," *Journal of Industrial Relations* 29 (4): 470–92.

Shash, A. (1993) "Factors considered in tendering decisions by top UK contractors," *Construction Management and Economics* 11: 111–18.

Thompson, H. and Tracy, J. (1993) "Building and Construction: Retaliation or Reform?" *Labour & Industry* 5 (1 and 2): 67–82.

Toner, P. (2000a) "Changes in industrial structure in the Australian construction industry: causes and implications," *The Economics and Labour Relations Review* 11 (2): 291–306.

Toner, P. (2000b) "Trade apprenticeships in the Australian construction industry," *Labour & Industry* 11 (2): 39–58.

Toner, P., Cooper, L., and Croce, N. (2001) *Industry Re-Structuring and Training in the Non-Residential and Engineering Construction Industry*, Employment Studies Centre, Newcastle: University of Newcastle.

Toner, P., Green, R., Croce, N., and Mills, B. (2001) "No case to answer: Productivity performance of the Australian construction industry," *The Economic and Labour Relations Review* 12 (1): 104–25.

Underhill, E. (1991) "Unions and contract workers in the New South Wales and Victorian building industries," in M. Bray and M. Taylor (eds), *The Other Side of Flexibility: Unions and Marginal Workers in Australia*, Australian Centre for Industrial Relations Teaching and Research, Monograph No. 3, Sydney: University of Sydney.

Underhill, E. and Kelly, D. (1993) "Eliminating traditional employment: Troubleshooters available in the building and Meat industries," *Journal of Industrial Relations* 35 (3): 398–423.

Underhill, E. and Worland, D. (2000) *Portable Sick Leave in the Victorian Building Industry: Managing Cumulative Employee Benefits in the Absence of Employment Security*, School of Management, Working Paper 1/2000, Melbourne: Victoria University of Technology.

Underhill, E., Worland, D., and Fitzpatrick, M. (1997) *Self-Employment in the Victorian Construction Industry: An Assessment of its Impact on Individual Workers and the Industry*, report prepared for the Redundancy Payment Central Fund Limited (Incolink), Melbourne: Incolink.

Wainwright, R. (2001) "Horses for courses: Bargaining in the construction industry," in *Ten Years of Enterprise Bargaining*, Proceedings of the conference at Newcastle, 3–4 May, Newcastle (New South Wales): Employment Studies Centre, University of Newcastle.

7 Spain
Spain down the low track

Justin Byrne and Marc van der Meer

INTRODUCTION

Since the early decades of the twentieth century, if not before, the construction industry has been the "engine" of the Spanish economy, one of the main driving forces of national production, wealth creation, and employment. In Spain, even more than in many other advanced economies, the crucial economic and social importance of the sector is summed up in the adage that "when the construction industry goes well, everything goes well." Given this, it is perhaps not surprising that the industry has also been at the forefront of broader trends in the organization of production, employment, and industrial relations. This was as true at the beginning of the twentieth century, when, for example, construction workers and employers led the way in interest group organization and conflict, as it has been recently, at the close of the century, when the industry led in developments such as productive decentralization, labor market flexibilization, and attempts to create new forms of industrial governance.[1]

These are the issues at the heart of this chapter, which is concerned with Spain during the period since 1975, when the death of dictator Francisco Franco y Bahamonde signaled the end of the authoritarian regime and paved the way for the transition to democracy. Democracy brought the legalization of independent trade unions and employers' organizations, the gradual elimination of the rigid corporatist labor legislation of the dictatorship, political and administrative decentralization, and, in time, the first steps towards the creation of a comprehensive welfare state and Spain's long-desired entry into the European Community, now the EU, in 1986. In other words, democratization meant the radical transformation of the political and institutional framework for economic activity and labor regulation. On the other hand, 1975 also saw the onset of a deep and protracted recession in the construction industry, which marked a major turning point in the evolution of the sector.

Democratic Spain has enjoyed long-term economic growth and rising prosperity. However, the Spanish economy has been unstable and still suffers from many of its historical problems. In comparison to its European partners, Spain has experienced high unemployment – averaging 17.5 per cent between 1990 and 1999, compared to the EU average of 9.7 per cent – and very low levels of both part-time and female employment. Since the early 1980s, intense labor market flexibilization has also brought extremely high levels of insecure, temporary employment; in 1998, almost 33 per cent of Spanish employees had one of a variety of fixed-term contracts, compared to the EU average of 12.8 per cent (UGT 2000: 7). Other traditional features of the economy include the existence of a large informal sector, or black-market economy. This, in turn, reflects the incapacity of the bureaucratic state to enforce its own laws, as well as the weakness of independent interest organizations and other intermediate bodies such as the official Labor Inspectorate. Non-enforcement of and non-compliance with the law are widespread across the economy and society with respect to, for example, taxation, employment conditions, and health and safety.

It is in this political and economic context that we locate the recent evolution of the Spanish construction industry. We argue that over the last two decades the Spanish construction industry has been through a process of profound re-regulation. In the product market, the contracting process in the public sector has been adapted to European standards on transparency, equality of opportunities, and competition, with a shift, in theory at least, away from pure competition on price (Fundación Encuentro 2000: 102–4). In the labor market, numerous new kinds of temporary contracts have been introduced since 1984, and the former provincial-level collective agreements have been replaced since 1992 by national-level agreements. Such agreements established sectoral arrangements and institutions for providing training and improving health and safety. At the same time, Spain first introduced stringent health and safety regulations in 1986 and has been applying European standards in this field since then.

Nonetheless, despite these developments, there is a wide gap between theory and practice. The current Spanish industrial order accepts substantial failure on the part of many agents in the construction industry to apply standards, resulting, among other things, in the fragmentation of the labor force and very high work-related injury rates. We suggest that these contradictory realities can be understood only in the context of the current division of labor between enterprises and the relatively weak performance of interest associations in the sector.

This chapter is divided into five sections. In the first, we trace the economic evolution of the construction industry over the last two decades. We then turn to the business structure, highlighting the causes and consequences of the core dynamics of productive concentration and fragmentation. In the third section, we examine labor market trends. In the fourth section, we move on to industrial relations, focusing on the creation of new mechanisms and institutions of industrial governance by the interest associations in the early 1990s. The shortcomings of these are highlighted most dramatically in the area of health and safety, which is discussed in the fifth section. Our conclusions signal some of the challenges facing the industry as it enters the twenty-first century.

ECONOMIC EVOLUTION: A CRUCIAL BUT TURBULENT SECTOR

While the construction industry is important in any national economy, its particular significance in Spain in recent decades is a consequence of the ongoing processes of urbanization, industrialization, and the expansion of mass tourism in the country, along with the demand these generate for building work of all kinds. Spain's particular stage and path of economic development explains why in comparative terms the construction industry accounts for a relatively larger share of total GDP than in most other EU member states.

Nonetheless, in Spain, as elsewhere, construction has historically been prone to particularly violent fluctuations in activity. The peaks and troughs in output are sharper than those in the economy as a whole (see Figure 7.1). Five clearly distinguishable phases of activity can be identified in the evolution of the industry over the last 40 years, as reflected both in the fluctuations of the gross added value of the sector and evolution of employment discussed below.

After almost two decades of economic autarky and stagnation in the wake of the Civil War (1936–9), the "technocratic" government appointed in 1957 opened up the Spanish economy at both the domestic and international level. The result was a period of furious expansion of the economy as a whole – with average GDP growth of 7.1 per cent annually between 1961 and 1974 – and, within it, of the construction industry (García Delgado and Jiménez 1999: 171).

This phase of expansion during the final years of the dictatorship was followed by a protracted recession in the sector from 1975 to 1985. In the context of the transition to democracy, political issues overshadowed the economic problems of an inefficient economy plunging into recession. Construction was particularly hard hit. While employment in the economy as a whole shrank by 17 per cent between 1975 and 1985, the number of construction workers dropped by 35 per cent, and unemployment in the industry rose from 11.5 per cent in 1977 to

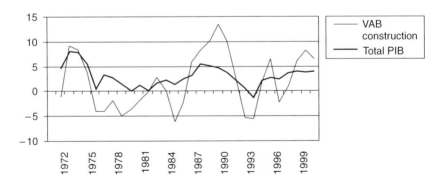

Figure 7.1 Annual percentage variation in total GDP and gross added value of the construction industry, 1971–2000.

Source: Contabilidad Nacional de España in CEOE 2001.

a staggering 32.8 per cent in 1985 (INE, *Encuesta de Población Activa* 1975–85; Carreras Yañez 1992: 233).

The upturn came after Spain's entry into the European Community in January 1986 and lasted into the next decade. Boosted by a major inflow of European funding, as well as the major construction projects undertaken for Spain's so-called "big year" in 1992 (with the Olympic Games in Barcelona, the Expo in Seville, and the European Capital of Culture in Madrid), the upturn brought intense job creation in the sector. Employment rose strongly between 1986 and 1990, when the industry finally pulled back-up to the levels of the mid-1970s (INE, *Encuesta de Población Activa* 1986–90).

In 1992, the sector again skidded into crisis amid the hangover after the year's fiesta. Investment, production and employment all fell from 1992 to 1994, as the sector shed 215,000 jobs and unemployment rose to over 25 per cent. Recovery began again in 1995. By 1998, outstripping the economy as a whole, production in the industry was rising by 5.7 per cent and employment by 9.6 per cent, riding on the back of increased activity fuelled by historically low interest rates, the demand accumulated during the recession, falling unemployment, higher wages for the growing number of those in employment, and the generally more optimistic climate (INE 1998; INE, *Encuesta de Población Activa* 1998). It was in this buoyant state that the industry entered the twenty-first century.

These violent cyclical fluctuations in construction not only have had obvious consequences for employment, but also have encouraged major changes in the productive structure of the industry and its labor market. The profound and protracted crisis in the late 1970s and early 1980s, in particular, constitutes the major turning point in the recent development of the Spanish construction sector. The industry that entered the recession in 1975 was very different from the one that emerged from the crisis in 1986, and many of the ongoing trends and characteristics of the sector can be traced to that watershed decade.

PRODUCTIVE STRUCTURE

As elsewhere, the Spanish construction industry stands out for the variety of type, size, and character of enterprise. Customarily, the main distinction in the sector is between the building and civil engineering branches. Building (*edificación*) covers the construction of housing and other private or public buildings and offices, as well as the maintenance of such buildings. Civil engineering, in turn, refers to public infrastructure and public works (*obra civil*).[2]

While the relative contribution of the different branches and subsectors to total output levels has varied considerably over time, on average new residential building, non-residential building, maintenance, and civil engineering each account for approximately one-quarter of the industry's total output.[3] In comparison to many other European countries, one distinctive feature of the industry in Spain is the relatively small size of the maintenance subsector's share of total output. This reflects the youth of much of the housing stock and can be expected to rise in the future. In terms of the breakdown of clients, the Spanish construction sector is unusually dependent on the State, with public investment accounting for an average 49 per cent of all activity between 1989 and 1998, fluctuating between 62 per cent and 33 per cent. The potential leverage this gives to the government administrations at various levels, particularly with respect to the larger companies who are their main direct suppliers, is immense, and has clear implications for any understanding of government regulation of the sector (FECOMA–CCOO 1993: 36; SEOPAN 1998; Fundación Encuentro 2000: 97).

In comparison to most of the other EU countries analyzed in this book, the Spanish product market in construction is relatively unregulated. Except in those specialist trades, such as electricians or gas installers, where the quality of work is of evident public interest, no formal professional qualifications are required of either workers or entrepreneurs. Like companies in any other sector, construction firms must register for fiscal and administrative purposes. An architect or engineer must supervise projects of any size, and for larger projects a health and safety plan is obligatory. However, firms are not obliged to meet any specific technical criteria in order to operate in the private sector, where the market operates free from either legally binding or customary rules on issues such as prices, standards, and competition. The contrast with the highly regulated product markets found, for example, in Germany could hardly be greater.

The public sector is subject to tighter regulation. All companies competing for public contracts of a value of over €120,000 must first obtain a certificate from the Ministry of the Economy and Finance entitling them to contract work with the administration. Using essentially financial and administrative criteria, such as economic resources, volume of work carried out in the past, and payment of all tax and social security dues, the Ministry classifies them into one of a number of categories entitling them to tender for public contracts worth up to a specified amount. Through Law 13/1995 on Contracts for the Public Administration, subsequently modified in Law 53/1999, Spain has incorporated into its national legislation certain European directives intended to ensure the transparency of tendering and contracting processes. However, critics complain that despite these mechanisms of regulation, and the shift from bidding to tendering for public contracts, implying the consideration of factors other than price, the failure to develop adequate systems to incorporate quality criteria into the contracting process has meant that, ultimately, price remains the key criterion when awarding public contracts (Fundación Encuentro 2000: 64–5).

Concentration and fragmentation

Over the course of the last two decades, the construction industry has undergone profound structural and organizational change, with concentration at the top and fragmentation at the bottom. The origin of both processes lies in the response of Spain's large construction companies to the crisis in the late 1970s and early 1980s. First, and perhaps most significantly, the largest companies opted to redefine their role in the production process, concentrating on management and coordinating functions while carrying out very little of the actual physical production themselves. This they increasingly subcontracted, often successively, to smaller firms and self-employed workers. The result has been a fundamental change in the nature of the large construction companies which "have turned essentially into mere service companies: they look for clients, contract companies that produce the product, and market it" (Miguélez 1990: 39).

This strategy of outsourcing has many benefits for companies in terms of the reduction of costs and risks. Since subcontracts are usually carried out for a fixed and closed price, subcontracting enables the principal contractors to control costs and lower prices. Principal contractors reduce the costs of work due to the competition among subcontractors and the tendency towards increasing special- ization. They also cut their own direct labor costs and virtually eliminate their own site personnel. Now, while Spain's 50-odd large companies – i.e., those with 500 or more employees – account for some 25 per cent of total output, they employ just 8 per cent of the workforce in the industry. Subcontracting also enables them to transfer the numerous risks of the construction process downward to the subcontractors who actually carry out the work. Another advantage of subcontracting for employers is that it hinders union development and action in the workplace. Unions have found it almost impossible to penetrate the world of the very small, transient, and mobile Spanish building firms, where worker representation is not required by law, and the threshold of collective organization remains beyond reach.

Subcontracting, of course, is neither a novelty in the industry nor exclusive to Spain. However, the unions argue that Spain now shows an unusually high – and they would argue, excessive – level of subcontracting. The official figures, though incomplete, show that, while in 1988 subcontracting accounted for 15.4 per cent of all operating earnings in construction, a decade later this figure had increased to 23 per cent. The evidence also points to a significant degree of chain, or multiple-level, subcontracting, where a principal contractor contracts the work, or part of it, to another company, which in turn, subcontracts the same unit of work or specific tasks to other companies, which in turn, subcontract it again, and so on.[4]

Intensive outsourcing through subcontracting has been accompanied by three other developments at the top of the industry. The first of these is the process of consolidation among the very largest construction companies. Although the level of concentration is still well below that of countries such as Sweden, France, and Germany, the tendency towards ever-larger enterprises is clear. The wave of mergers among Spain's largest companies in the 1990s meant that by the end of the century the "big dozen" of the mid-1980s had been whittled down to the "big five": *Fomento de Construcciones y Contratas* (FCC), Dragados, Acciona, ACS, and Ferrovial.[5]

These very large companies have also led the way in two other tendencies in the industry: diversification and internationalization. In both cases, the driving force is the desire to reduce their dependency on their core business (construction) in a highly volatile market (Spain). Diversification has taken the largest companies into a variety of other sectors, from the production of building materials to tele-communications, the operation of seaports and airports, and the provision of municipal services and utilities. Many of these activities have been developed outside Spain, with diversification going hand in hand with internationalization, in Latin America, North Africa, and the EU (SEOPAN 1998: 13–14).

Diversification and internationalization are mainly, if not exclusively, charac-teristics of the 50-odd largest companies with 500 or more employees and operat-ing at the national level. Below these are the more than 200 medium-large companies (100–500 employees). Responsible for around 7 per cent of total output and 5 per cent of employment, these medium-large companies also work on both civil engineering and building projects, and operate mainly as principal contractors at the regional level, with their competitive advantage deriving from their greater understanding of the local conditions and their contacts with regional and local administrations. They occupy an intermediate position in the subcontracting chain, in that, above all in the case of some highly specialized companies, they carry out some work for the larger firms, while also subcontracting out work themselves to other, smaller firms.

This is also the case for the approximately 8000 medium-sized enterprises (20–99 employees), which account for about 15 per cent of the total output and employ 10 per cent of the workforce. These companies tend to be very highly specialized in terms of geographical scope and/or specialty, as well as in terms of product market and client. Focused more toward building and maintenance than civil engineering, and toward the private rather than the public sector, they perform on average some 20 per cent of their activity as subcontractors.

The pyramidal shape of the business structure of the Spanish building industry flattens out dramatically below this point, as the vast majority of enterprises in the sector are small enterprises – in many cases micro-enterprises. The roughly 137,000 small building firms – those with fewer than 20 employees – account for almost 95 per cent of all enterprises in the industry. Responsible for about 34 per cent of the sector's total output, and employing some 41 per cent of the work-force, they also draw some 20 per cent of their work from subcontracting relationships with other, generally larger firms. This is much less important in the case of the small, independent, all-round builders who work on small-scale building projects and repair and maintenance work, than in the other two char-acteristic types of small firm: specialist companies devoted to installations and finishing (e.g., plaster, tiling, and floor layers) and labor-only subcontractors, whose function is exclusively to supply labor to other firms (Miguélez 1990: 30–3).

At the very bottom of the pyramid of the business structure of the industry are the self-employed, who in fact constitute the largest single category of "enter-prises" in the sector. As individual self-employed "workers" are themselves allowed, under the Spanish employment and tax regimes, to employ one or two assistants, their overall share of employment (380,000 in 1994) is higher than the actual number of self-employed (184,000). Self-employed workers with no employees account for some 45 per cent of all construction firms, while those with 1–2 employees (i.e., skilled workers working with one or two assistants) represent another 25 per cent. Most of the single self-employed and these microenterprises work in the building subsector and comprise bricklayers, painters, tilers, plasterers, and other tradesmen carrying out installation, maintenance, and repair work. They often work as subcontractors; in this sense the increase in the proportion of self-employed with employees as a proportion of the total workforce in the sector – according to the Active Population Survey (*Encuesta de la Población Activa*, or EPA), up from 4.6 per cent in 1988 to 8.1 per cent in 1999 – reflects the expansion of outsourcing and chain subcontracting in the sector (Fundación Encuentro 2000: 126). At the end of this chain, as we shall see below, many of the one-person enterprises are so-called *falsos autónomos*, more or less willingly "false self-employed" (Miguélez 1990: 50–2; Babiano 1993: 177).

Internal organization of firms

The evolution of the business structure of the Spanish construction industry shows, therefore, divergent tendencies: concentration and diversification at the top in marked contrast to fragmentation and specialization at the bottom. At the top, the largest companies now form part of large, highly diverse international groups in which construction is just one of many activities, and one of which they themselves do only a small part. At the bottom, more and more workers are employed by medium-sized to small companies and microbusinesses, created to supply the products and services required by firms higher up the chain.

The core dynamic linking these two tendencies is outsourcing through subcontracting, a longtime reality in the sector but one which has expanded considerably over the last two decades, not least through the spread of chain subcontracting. Extensive multiple-level subcontracting allows the largest companies to move into higher-added-value, more risk-averse areas of activity. They have now become huge publicly floated corporations, with a hierarchical internal structure, a highly skilled labor force of essentially technical and managerial staff, and relatively professional human resources policies and practices. These companies have virtually no onsite operatives, as the actual physical work of construction has been decentralized to subcontracting companies.

For the subcontractors down the chain, a process of specialization and fragmentation is taking place. In terms of their internal structure and organization, very small firms are notable for their simplicity. They take advantage of the low start-up costs in building, the absence of any technical or professional requirements for those wishing to set up building or labor-supply firms, and the State's failure to regulate or supervise their activities. Often comprising just two or three employees, who may well be members of the same family, in which the father, typically, is also the boss (*padre-patron*), they often have no real internal structure or management or administrative personnel, beyond perhaps the part-time support of another, often female member of the family (Prieto 1991: 200). These micro-businesses function both internally and externally through mainly informal networks and relationships. Although such relations and networks can be stable and long-term, within them these firms tend to operate with a very short-term perspective.

The spread of chain subcontracting is the key dynamic in the Spanish construction industry, and one that is at the heart of many of its problems, including the quality of the final product, of employment, of working conditions, and of the workforce itself.

THE LABOR MARKET

The construction industry is a crucial source of employment in Spain. During the 1990s, it averaged around 10 per cent of the total labor market in Spain, compared to 62 per cent in services, 20 per cent in industry, and 8 per cent in agriculture. It is, however, a particularly unstable source of employment, as activity and employment in the industry fluctuate more sharply than in other sectors, dropping deeper in periods of crisis and rising particularly intensely during upturns in the economy (Figure 7.2).[6]

As we review employment over the last three decades, we see a slump in the second half of the 1970s, followed by a dive in the early 1980s, when registered unemployment in the sector hit nearly 33 per cent in 1984 compared to 23 per cent in the economy as whole. Recovery began in 1985, but lasted only until 1991; between 1992 and 1994 the industry lost 200,000 jobs. Since 1995, the total workforce in the sector has been growing steadily, passing the 1991 record of almost 1,300,000 in 1998, and topping 1,500,000 in 2000.

As in the sector as a whole, elements of continuity and change can be seen in the structure of employment in the construction industry over the course of the last few decades. In terms of continuities, the labor force is still almost entirely male, full time, and comparatively poorly trained and educated.

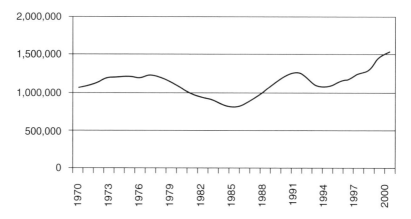

Figure 7.2 Employment in the Spanish construction industry, 1971–2000.
Source: INE, *Encuesta de Población Activa*, 1971–2000.

In 1999, women accounted for only 5 per cent of the total workforce in the sec-tor and were confined almost exclusively to administrative and technical posts; onsite "their presence is purely anecdotal" (Fundación Encuentro 2000: 117). In the same year, just 1.4 per cent of all construction workers had part-time contracts, compared to 8.3 per cent in the economy as a whole. The vast majority of part-timers were women; among male technical and management staff, as well as site operatives, overtime rather than part-time work is the norm.

Inadequate training

Compared to the rest of the economy, the construction industry stands out as a refuge for workers with limited formal education or training. Over 42 per cent of the workforce either has no formal educational qualifications or completed only compulsory schooling. Only 11.7 per cent of construction workers have any kind of formal professional qualification from the state vocational training system, in which only 1 per cent of all registered students in 1996–7 were taking courses in building (MCA–UGT 2000: 41–2, 73). As has historically been the case, most workers still enter the industry through informal channels – i.e., family or personal networks – and training still takes places almost exclusively on the job. Those entering or already working in the higher technical professions bene-fit from greater provision for training in high schools and universities, which offer courses and qualifications in engineering, architecture, technical drawing, project management, and so on, as well as from training offered by the larger companies and their associated research institutes. In contrast, non-professional, non-technical workers' skills and performance are sanctioned by customary and informal norms and evaluation rather than by official certificate.

The low levels of education and formal training found among Spanish con-struction workers can be traced back to a number of interrelated factors. These include the traditional low status of construction work, the obstacles to training posed by the transient nature of both production and employment, the limited interest in, or capacity for, training shown by the small companies employing most workers in the industry, and inadequate provision for training by both the public sector and the industry itself. In this respect, observers argue that the construction industry suffers from many of the problems inherent in Spain's state-run vocational and professional training systems: the emphasis placed by the Spanish educational system, parents, and potential entrants into the sector on general education and academic development; the actual quality of the training provided; the excessively theoretical leaning of many courses, and the predomi-nance, in construction, of courses for technical staff rather than site operatives; and more generally, the failure to adapt to the real needs and conditions of the economy. As discussed below, the occupational training courses run since 1993 by the bipartite (joint labor–management) Labor Foundation for the Construc-tion Industry (*Fundación Laboral de la Construcción*, or FLC), share a number of these defects and have had no significant impact on training levels among site operatives.

Temporary employment and self-employment

Alongside these continuities in the labor market, two highly significant trends can be identified in the structure of employment in the sector: the persistent expansion of fixed-term or temporary employment, and the increase in the level of self-employment in the mid-1980s.

Construction work has always been associated with temporary work, reflecting the fluctuating demand for labor in the industry as a whole as well as on any specific project. Even under the very rigid labor market regulations of the Franco dictatorship, when security was one of the few carrots the regime offered to the working class, temporary employment was effectively allowed in the construction industry in the shape of "permanent contracts for the duration of the job." In contrast to administrative and technical staff, site operatives experienced this as the predominant form of employment contract and became members of the permanent workforce only after they had worked for the same company for a period of 2 – and, from 1970, 4 – years (Ruiz and Babiano 1993: 106, 178–80).

While, therefore, the Spanish construction industry has always had an unusually high level of temporary employment, this increased dramatically during the late 1980s and early 1990s (Figure 7.3). When the industry emerged from its decade-long slump in 1985, most companies did not replace the permanent staff lost in the recession, but rather gave temporary contracts to virtually all new entrants. These temporary contracts, with no real prospect of becoming permanent, were permitted under the labor market reforms the Socialist government introduced from 1984 onwards as part of their bid to boost employment by "flexibilizing at the margins" (García de Polavieja 1998). Since 1984, the contractual form *par excellence* in the sector has been the contract for a specific task or service (*contrato por obra o servicio*). This has no minimum duration – most of those in which the length is specified last less than one month – and implies no severance pay in case of non-renewal or dismissal. The proportion of temporary workers in

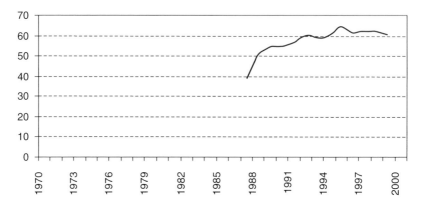

Figure 7.3 Temporary workers (as percentage of all construction wage earners), 1987–99.

Source: INE, *Encuesta de Población Activa*, 1987–99.

the sector doubled between 1987 (when data became available) and 1991, and continued to rise over the course of the 1990s. According to the provisional results of the EPA for 1999, 61.7 per cent of construction employees had a temporary contract, while a full 97 per cent of all new contracts in the industry were temporary. These figures compare with the temporary work rate of 32.7 per cent in the Spanish economy as a whole. Moreover, in 1996 the EU average of 18.1 per cent temporary employment in construction was over 40 points below the Spanish level (Fundación Encuentro 2000: 118–20; MCA–UGT 2000: 35–9). While, therefore, in construction the temporalization of employment predates the ongoing process of labor market reforms initiated in 1984, these legal changes have undoubtedly intensified labor market dualization, with the proportion of temporary work rising significantly and consistently as one moves down the skill hierarchy in the industry (FECOMA–CCOO 1993).

The second significant tendency in the structure of the labor market has been the expansion of self-employment (Figure 7.4). Once again, this is a process that can be traced back to the watershed decade, 1975–85, when the sector shed nearly half a million jobs, but gained 70,000 self-employed workers. This sent the self-employment rate in the sector up from around 16 per cent in 1976 to 26 per cent in 1985. While this figure fell again in the second half of the 1980s, since then it has fluctuated between 17 and 23 per cent. In the case of the self-employed workers without employees, there is a clear if imperfect correlation between the self-employment rate and the level of activity in the sector: self-employment rises during periods of crisis and drops again during periods of expansion. This suggests that during downturns employers take advantage of the labor market conditions to take on workers as self-employed rather than wage labor, thereby saving themselves the social insurance costs associated with salaried workers. Many of those taken on in these conditions are considered to be *falsos autónomos*, salaried workers in all but name and rights, whose absolute job insecurity makes them particularly vulnerable to the risk of self-exploitation, work

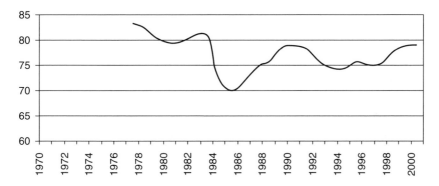

Figure 7.4 Wage earners (as percentage of total workforce in construction), 1977–2000. Source: INE, *Encuesta de Población Activa*, 1971–2000.

injuries, and unemployment (Miguélez 1990: 50–2; Babiano 1993: 175–8; Fundación Encuentro 2000: 126–7).

There is another segment of the labor market that entails even greater risks and worse conditions for construction workers: the informal sector. Although its numbers are inevitably almost impossible to assess, a study on Madrid published in 1994 estimated that perhaps 5 per cent of work in large firms and 10–50 per cent of labor in small firms is carried out through informal circuits. In addition, the informal sector accounts for as much as 50 per cent of all subcontracting, and 50–80 per cent of self-employment (CCOO–Madrid 1994: 221–31). This sector includes an unknown number of immigrant workers working illegally, especially in the larger cities like Madrid and Barcelona and in tourism-related projects on the Mediterranean coast. Illegal migrants in Spain certainly outnumber legal migrants, who, even after the collective legalization of formerly illegal immigrants in 1991 and again in 1996, still numbered under 5000 in 1998 (Colectivo Ioé 1997; Fundación Encuentro 2000: 129–30).

To sum up, employment in the construction industry, which has always played an important role in the Spanish economy, is now at an all-time high of one and a half million workers. The majority of these are male, full-time workers, a large proportion of whom has few formal qualifications, even as compared to other manual occupations. The upturn in employment has had no noticeable impact on the quality of employment. The majority of site operatives and virtually all new entrants into the sector now have temporary contracts lasting only for the duration of the job or the task on which they have been contracted to work. They work alongside a significant proportion of self-employed workers, many of whom are *falsos autónomos*, whose relationship to their client is essentially that of a waged worker to an employer. Symptomatic of the increasing fragmentation of the productive structure of the construction industry, the insecurity of employment is also linked to other problematic features of the industry identified here. Insecure employment operates, for example, against both training and union organization and supervision of the workplace, while encouraging workers to accept an excessive pace of work and overtime in order to show willingness and to make the most from work while it lasts.

THE CORPORATIST ILLUSION: SOCIAL PARTNERS AND LABOR RELATIONS

The social partners

Free, independent unions and employers' organizations are recent phenomena in Spain. Both were prohibited during the Franco dictatorship, when membership in the "vertical" syndicates set up by the regime was obligatory for both workers and employers. While clandestine trade unions existed during the dictatorship and became increasingly active in the 1960s, independent interest organizations were not legalized until April 1977 in the midst of the transition to democracy.

In the construction industry, the representation of workers is essentially divided between the relevant sectoral federations of the two main nationwide trade union confederations, *Comisiones Obreras* (CCOO), historically associated with the Communist Party, and the Socialist-led *Union General de Trabajadores* (UGT). Although competing for members and representation of labor in the sector, the two national confederations have made unity of action the norm since the 1980s. This has also been the case in construction, although relations between the CCOO and UGT construction federations remain tense. Even within the context of the low levels of union membership across the Spanish economy as a whole, the construction industry stands out for the weakness of union organizations: in 1997 CCOO claimed a national membership of just over 42,000 and the UGT claimed nearly 38,000. These figures give a union affiliation rate of just 10.7 per cent in 1997, compared to the economy-wide average of 17.8 per cent (van der Meer 2000: 573–603). The trade unions in the sector, therefore, are relatively recent, small, and underdeveloped organizations. They are heavily dependent on the financial support of the State and have only a limited presence in the workplace, chiefly within small firms and on small construction sites.

The employers' organizations in the industry face similar problems and challenges. They too are still in their adolescence, with relatively small memberships and a limited level of organizational development. At the national level, there is one peak-level employers' organization, the National Building Confederation (*Confederación Nacional de la Construcción*, or CNC). Since its foundation in 1977, the CNC has grouped together most nationwide employers' organizations from the different subsectors of the industry, as well as the regional and provincial employers' federations. In all, the CNC claims to represent a quarter of the approximately 270,000 firms in the construction industry, but, since these are the larger firms, its members directly or indirectly provide employment for a significant proportion of the total workforce in the industry. The dominant force within the CNC is the *Asociación de Empresas Constructoras de Ámbito Nacional* (SEOPAN 1998), the association of the very largest construction companies. SEOPAN currently has a membership of under 30 companies but claims that these are responsible for 22 per cent of all construction work carried out in Spain (SEOPAN 1999). The other major forces within the CNC are the regional federations, which retain considerable autonomy within the organization, a fact that, along with its low overall membership, further undermines the CNC's capacity to represent and ensure the collective discipline of employers in the industry.

Rather surprisingly, however, over the last 10 years these representative organizations in the construction industry have been able to develop national-level collective bargaining for the first time, as well as to create the first institutionalized joint labor–management councils of industrial governance in Spain.

A new system of industrial governance

Following a number of regional initiatives in the late 1980s, especially those in Asturias and Catalonia, and thanks to the bargaining power the construction unions enjoyed in the rush to complete the major showpiece projects for Spain's big year in 1992, as well as the political clout the UGT union confederation's *Federación Estatal de Madera, Construcción y Afines* (FEMCA) enjoyed with the Socialist government, in 1992 the unions were finally able to convince the CNC to sign an agreement creating a new bargaining structure at the national level (van der Meer 1998: 245–68). This *Convenio General del Sector de la Construcción* marked a key turning point in industrial relations in the sector: first, in that it radically altered and rationalized the bargaining structure in the sector, and second, because it provided for the creation of the *Fundación Laboral de la Construcción* (FLC), the first bipartite institution of its type in any Spanish industry.

The signing of the 1992 *Convenio General del Sector de la Construcción* undoubtedly represented a major step forward in the rationalization and modernization of labor relations in the industry. Replacing the previously fragmented, provincial-based bargaining process, the *Convenio General* established national-level bargaining as the main level of wage setting, and it determined virtually all other aspects of labor relations, with the corresponding savings in bargaining resources and improved sector-wide coordination of conditions and policies. Negotiations still do take place at the provincial level, but these are to decide only such issues as the distribution of working hours and regional holiday time over the course of the year, some travel and food supplements, and the organization of the regional branch of the FLC. At the same time the *Convenio General* substituted the old Francoist labor ordinance with collective labor regulations better suited to the needs of the industry (Valdés dal-Ré 1993). Under the general agreement, wages and hours have been the object of specific, generally annual, national-level bargaining between the social partners. Like the *Convenio General*, any such *Acuerdo Sectorial Nacional de la Construcción* is extended by *erga omnes* (Latin: "in relation to everyone") clauses, which make the agreements binding even on unorganized firms. These agreements refer exclusively to wage levels – i.e., setting minimum wage levels, with a guarantee clause applicable in event of higher-than-anticipated inflation – and the duration of the annual working year.

Despite the undeniable importance of the consolidation of this new structure of collective bargaining, it should be emphasized that the practical impact of the agreements is much more limited than it at first appears. All the evidence suggests that the terms of agreements are honored as much in the breach as in the observance. Statistics point to seasonal and business cycle specific fluctuations in wages, with collective agreements serving essentially as a base line. Hence, workers will often secure higher-than-official minimum wage rates in good times, while in bad times unpaid overtime is common and real wages may fall below the official rate. This reflects the much more general problem of adherence to and enforcement of the terms of any collective agreements. Given the unions' weak presence in the workplace, it is almost impossible for them to enforce agreements, especially in small firms and sites. With only a blurry boundary between the formal and informal economy, non-compliance with tax, social security, and health and safety regulations and standards is also rife, and loudly denounced by the unions.

The Fundación Laboral de la Construcción (FLC)

In the second half of the 1990s, the unions also began to show increasing dissatisfaction with the other major product of the original 1992 *Convenio General*, the FLC. The FLC is not only a unique institution in the Spanish context, but also an unprecedented development for the construction industry. A national-level institutionalized bipartite council of industrial governance, the FLC was entrusted with three main tasks: (1) to provide vocational training for workers in the sector; (2) to study and promote measures to improve health and safety; and (3) to promote craftsmanship through the distribution and supervision of a certificate of seniority (*cartilla profesional*). Although a national organization, and governed by a national-level governing board, in operational terms the FLC is decentralized to the regional level, where decisions are made by bipartite regional committees. The FLC is funded, first, by an obligatory employers' quota (set at 0.05 per cent of the annual salary base in 1998, plus a special one-time payment of 0.1 per cent to fund the restructuring process), and second, by the grants and subsidies it receives from different levels of the government for its activities and for providing training programs.[7]

The record of the FLC since 1992 has been very mixed. In terms of its three original goals, it has probably been most effective with respect to the first, training. In response to the evident failure of the excessively theoretical, inflexible, and low-status state system to meet the labor needs of the industry, the FLC was charged with reviewing and designing vocational training programs. It has also provided training directly, organizing hundreds of courses each year for unemployed workers and providing continuous training for employed workers. However, its training activities have been in decline since 1997, when the FLC ran into economic difficulties due to a sharp decline in inflows of public funding for training. At the same time, most courses offered by the FLC have been for administrative and technical personnel rather than for site operatives, who are those most in need of greater skills and qualifications.

In terms of its second goal, promoting health and safety at work, the FLC has carried out research and produced a wide range of handbooks and other materials. In addition, it has organized numerous seminars and training courses in health and safety; in 1998, health and safety topics accounted for nearly a quarter of all the continuous training courses run by the FLC. However, it has scarcely fulfilled the more active role in accident prevention assigned to it in the "Addendum on Work Safety" attached to the second *Convenio General* in 1998. Significantly, only in January 2001 did the social partners finally agree on the design and functions of the joint "specific body" that, according to the 1998 agreement, would be responsible for accident prevention in small companies without worker safety delegates.

Finally, no progress has been made in the third main task assigned to the FLC in 1993, the introduction of a certificate of seniority, which was intended to eliminate trial periods for workers employed with "permanent contracts for the duration of the job."

The FLC was also intended to have a positive influence on the climate of labor relations in the industry, by providing a forum for a trust-enhancing dialog between the social partners. Employers and unions are certainly in much closer contact now than a decade ago, and relations now embrace cooperation as well as confrontation. While both employers and labor leaders continue to see this as an intangible benefit of the institution, in recent years the trade unions have become increasingly frustrated with the performance of the FLC. They argue that it is suffering from a crisis of resources, effectiveness, leadership, and vision. The need for unanimity in decision-making within the FLC and its dependence on employers' contributions mean that its development has inevitably been undermined by the employers' alleged lack of commitment to engaging in cooperative relations and initiatives.[8] In this context, the unions' continued willingness to work within the FLC could be seen as an indication of their own weakness and the absence of other viable strategies, rather than as an endorsement of the existing framework of labor relations in the industry. Union dissatisfaction with results of the new system of industrial governance is particularly acute with respect to health and safety. The ever-rising injury rate in the sector constitutes a particularly clear-cut manifestation of the failure of neo-corporatist arrangements developed in the sector in the 1990s.

THE DARK SIDE OF THE BOOM: DANGER AT WORK

Not only does Spain have the worst record for accidents within the EU, but also its construction industry stands out above all others. In 1998, the last year for which complete figures are available, the incidence of work injuries stood at 17.5 per cent in construction, compared to 10.7 per cent in industry, and 4.5 and 3.8 per cent in the service and agricultural sectors respectively (MTAS 1998). As can be seen in Figure 7.5, injury rates have been rising consistently since the mid-1980s and even more dramatically in recent years. Thus, the incidence rate of 13.9 per cent in 1990 increased to 15.2 per cent in 1995, 16.4 per cent in 1997, and 17.6 per cent in 1998. Moreover, the provisional figures for 1999 show a further 25 per cent increase in the total number of incidents.[9]

It should be emphasized that this persistent increase in the reported injury rate has taken place in a context that *a priori* would appear to favor improved work safety. The legal framework both for the economy as a whole and for the construction industry in particular has been strengthened considerably since the mid-1980s, not least in order to bring the law in line with European requirements. This process is now complete and has meant, for example, that all but the smallest sites are obliged to have health and safety plans, committees, and worker representatives, as well as to comply with an increasingly rigorous series of safety norms and standards.[10] Moreover, since 1996, in particular, the previously light-hand of the Labor Inspectorate has fallen somewhat more heavily on the sector, and the government has proudly announced a series of government "emergency plans" on safety in the construction industry. At the same time, the

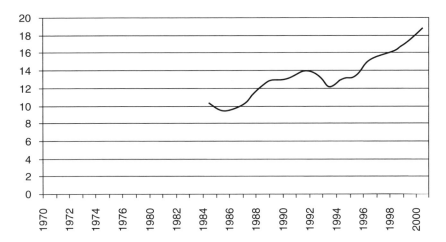

Figure 7.5 Incidence of work-related injuries (and mortalities), incidents per year per hundred workers, 1983–99.

Source: Elaborated by the authors from the following sources: MTAS 1998, FECOMA–CCOO 1993; and provisional data for 1999 obtained from staff members of the Instituto Nacional de Higiene y Seguridad en el Trabajo, Madrid, September 2000.

social partners have launched numerous initiatives on health and safety involving the FLC. Nevertheless, neither tougher *de jure* state regulation nor self-regulation by the industry itself has halted the upward spiral in the industry's injury rate.[11]

Existing regulations and systems of industrial governance have proved, therefore, inadequate to counter the negative safety implications of changes in the structure, internal organization, and employment and working conditions of the construction industry. A number of developments have particularly obvious implications for work safety.

Perhaps, the most important of these is productive decentralization through subcontracting. Subcontracting, first, increases the difficulties involved in coordinating work and ensuring onsite safety, as employees of various companies, each with its own priorities and safety systems, will often be working simultaneously. Through subcontracting, the absence of a single framework of labor relations also means that obligations and rights tend to become blurred; while the principal contractor is legally responsible for ensuring compliance with safety norms and plans on the site, work is actually supervised and organized by subcontractors and carried out by workers who often do not know from whom they can demand safe working conditions or to whom they can report dangers. Subcontracting, which has been associated with the division of units of work into ever smaller and more specialized tasks, has also favored increasing atomization of the productive structure in a sector already characterized by its fragmentation. As is well documented in both construction and other branches of the economy, there is a positive relationship between company size and safety. This is due, among other factors, to the reduced financial and technical resources of smaller companies, which leads to a lack of capacity to implement safety plans, train employees, or sustain health and safety committees. Finally, the relationship between competitive subcontracting and accidents springs from the way in which the former intensifies productivist pressure. All these factors help explain why, according to the unions, 95 per cent of serious and fatal injuries involve workers employed by subcontractors (CCOO 2000).

The spread and intensification of subcontracting has gone hand in hand with a second key variable in any explanation for the rising accident rates in the sector: the ever-greater proportion of temporary workers. Temporary work makes it less likely that workers will gain the training and/or experience required to work safely in a dangerous environment. Temporary workers are also more likely to take risks, in that they are under greater pressure to demonstrate their productivity and in a weaker position to demand safe working conditions.

Structural tendencies in the organization of production and employment appear, therefore, to be undermining the potentially positive impact of greater regulation and self-regulation in the Spanish building industry. In fact, the issue of health and safety highlights the limitations of the neo-corporatist tendencies seen in the industry in the 1990s. Safety has been the main issue at stake in all the major industrial disputes in the sector in recent years, including three national general strikes (in 1998, 2000, and 2001), as well as an underlying cause of tension between the social partners. This was apparent in the social partners' statements to the Spanish Senate's major enquiry into workplace accidents in 1995 (Senado de España 1995), as well as in the employers' rejection of the unions' 1999 Popular Legal Initiative (*Iniciativa Legislativa Popular*, or ILP) to regulate and restrict subcontracting in the sector. Intended to tackle two of the underlying causes of work accidents – uncontrolled subcontracting and temporary employment – the ILP sought to set up tighter control of firms working as subcontractors, prohibit chain subcontracting, restrict the volume and type of subcontracted work, and ensure minimum levels of permanent employment by subcontractors. In November 2000, the conservative majority in the Spanish parliament (*Las Cortes Generales*) rejected the bill on the grounds that it represented an undesirable and unnecessary restriction on private initiative and the market, a decision that was welcomed by employers. Their reaction was hardly surprising, since the unions' proposals went directly against the subcontracting, pay, and employment policies that Spanish construction companies have been implementing for 15 years or more. Moreover, while employers continue to make declarations of good will and to defend the virtues of self-regulation, the unions are clearly losing confidence in this (Byrne 2000).

CONCLUSIONS

The Spanish construction industry has much to be happy about. Twenty years after the transition to democracy, the market and regulations have substantially been adapted to European standards. The onus of collective bargaining in the sector has shifted from the provincial to the national level. The industry is currently enjoying the longest-lasting period of growth in recent history. At the top, consolidation has brought a number of very large, profitable, internationally oriented, diversified groups. Below them, the boom has brought business and profits for firms of all sizes across the sector, and employment opportunities and comparatively high earnings for a record number of construction workers.

However, the industry is beset with a number of very serious, interrelated problems: the fragmentation of the productive structure, encouraged by the spread of chain subcontracting; the shortage of skilled workers in a growing number of trades, and the low-level of, and poor provision for, training; the very high rates of workers on fixed-term contracts and the tendency towards labor market segmentation, above all for immigrant workers; and the very high, still rising injury rates in the industry. Although mentioned only in passing here, there is also an underlying concern about the quality of construction work carried out in Spain. The Spanish construction industry seems well ensconced on the low-wage, low-skilled path signaled by Peter Philips and Gerhard Bosch in their introduction to this volume.

We argue that structural change, labor market flexibilization, and the lack of effective enforcement of contractual and labor legislation by the State, are the forces that have sent the industry down this path. Concerted intervention and cooperation by unions, employers, and the State would be required to put the industry on the high-wage, high-skilled path. The current situation does not appear very positive in this respect. The trade unions lack the substantial critical mass required for collective action. Among employers, the largest enterprises dominate the representation of the sector, while smaller companies have very little policy-making influence or access to the collective benefits provided in the fields of training and health and safety. Both social partners, but above all the trade unions, have expressed increasingly negative evaluations of the results of the existing neo-corporatist arrangements. According to the unions, responsibility for this lies almost exclusively with employers, and their actions reflect a lack of commitment to the principles of social partnership.

There is much evidence to support these charges. On the other hand, it remains to be seen how far the State is willing to intervene for the benefit of the industry, and whether the government's verbal commitment to promoting stable employment, cutting accident rates, and improving provisions for training will be matched by tangible action. In this respect, the current Partido Popular (PP) government's failure to allow parliamentary discussion of the union's proposed bill on the regulation of subcontracting or to support a cross-party bill on accident prevention does not bode well for the future (Byrne 2000). For there is no sign that the sector's problems are easing, or that, left to its own devices, the Spanish building industry can resolve the major challenges it faces.

NOTES

1 For the evolution of the industry since the mid-nineteenth century, see Byrne 1993 and 1998, Ruiz and Babiano 1993; for developments after the transition to democracy, see van der Meer 1998.
2 In the North American context, civil engineering contractors are often referred to as "heavy-and-highway contractors."
3 This analysis of the productive structure of the industry draws on a variety of scattered, often contradictory sources. The data on the breakdown of firms by number of employees comes from the official *El Directorio Central de Empresas* (see INE 1999).
4 This is what Mark Harvey terms "flexibilization with fragmentation" in his insightful typology of subcontracting arrangements found in Britain (see the United Kingdom chapter in this book).
5 According to 1997 data, the country's largest company, FCC, ranked just twelfth among the largest construction companies in the EU (UGT 2000). Figures are for the production and workforce of the 20 largest companies from 1980 to 1998 as listed in *El País*, 13 February 2000.
6 For the data on which the following analysis of the labor market is based, see in particular Fundación Encuentro 2000: 114–22; and MCA–UGT 2000: 32–42.
7 For more information about the FLC and its development, see Fundación Encuentro 2000: 144–8; and MCA–UGT 2000: 98–102.
8 Interviews carried out by Justin Byrne with representatives of the CCOO and UGT construction workers' federations, June–July 1999 and July–September 2000.
9 International comparisons of workplace health and safety data are notoriously difficult, due to the use of different definitions, data collection, and processing methods. Spain, for example, uses a relatively broad definition of workplace "accidents." These comprise all work-related accidents leading to the loss of working time, including, for example, trip-related injuries and cases of work-related medical conditions that cause time off work. At the same time, however, the statistics pertain to incidents involving only workers covered by the state insurance funds, and not, therefore, the self-employed or workers employed in the informal economy. Moreover, injuries are classified as light or serious on strictly medical criteria, irrespective of the number of days lost from work.
10 Especially, decree 555/1986.
11 It is possible that the rising injury rates, pushed up above all by the increase in the number of "light injuries," may in fact, in part at least, be due to tighter regulation and supervision. Significantly, however, this is not an argument used by the actors in the sector. While they do admit the possible impact of fraudulent reports of "light accidents"(without explaining why these might have increased), even the most complacent employer and official representatives do not suggest that higher reporting rates are responsible for the increase in the figures. See, for example, the Director of the National Institute for Workplace Health and Safety, Leodegario Fernández Sánchez (2001).

REFERENCES

Asociación de Empresas Constructoras de Ámbito Nacional (SEOPAN) (1998) *Informe anual*, Madrid: Agrupación Nacional de Contratistas de Obras Públicas (ANCOP).
Asociación de Empresas Constructoras de Ámbito Nacional (SEOPAN) (1999) *Folleto informativo*, Madrid: SEOPAN.

Babiano, J. (1993) "La mano de obra en la construcción madrileña: quince años de transformaciones (1975–1990)," in D. Ruíz and J. Babiano (eds), *Los trabajadores de la construcción en el Madrid del Siglo XX*, Madrid: Fundación Primero de Mayo.

Byrne, J. (1993) "La construcción durante el primer tercio del siglo XX," in D. Ruíz and J. Babiano (eds), *Los trabajadores de la construcción en el Madrid del Siglo XX*, Madrid: Fundación Primero de Mayo.

Byrne, J. (1998) *The Bricklayers of Madrid: Work, Organisation and Conflict, 1870–1914*, Ph.D. thesis, Florence: European Institute.

Byrne, J. (2000) "'Nos va la vida': La Iniciativa Legislativa Popular para una ley reguladora de la subcontratación en el sector de la construcción," *Sociología del Trabajo* 45 (Autumn): 93–107.

Carreras Yañez, J. L. (1992) "Perspectivas de la Construcción en los años 90," *Papeles de la Economía Española* 50: 210–38.

Colectivo Ioé (1997) *Inmigración y trabajo. Trabajadores inmigrantes en el sector de la construcción*, unpublished working paper, Madrid: Instituto de Migraciones y Servicios Sociales (IMSERSO).

Comisiones Obreras (CCOO) (1994) *El sistema de formación profesional en España*, Madrid: Ediciones GPS.

Comisiones Obreras (CCOO) (2000) *Gaceta Sindical*, June, Madrid: Confederación Sindical de CCOO.

Comisiones Obreras–Madrid (CCOO–Madrid) (1994) *Economía Sumergida*, Madrid: Ediciones GPS.

Confederación Española de Organizaciones Empresariales (CEOE) (2001) *Informe Socioeconómico* (June), Madrid: CEOE.

Federación Estatal de Construcción, Madera y Afines de Comisiones Obreras (FECOMA–CCOO) (1993) *Construcción: crisis y expectativas: análisis de la estructura interna del sector*, Madrid: Ediciones GPS.

Federación Estatal de Metal, Construcción y Afines de la Union General de Trabajadores (MCA–UGT) (2000) *Dificultades de accesoa la formación continua de los trabajadores del sector de la construcción*, Madrid: MCA/UGT.

Fernández Sánchez, L. (2001) "¿Cómo interpretar las estadísticas de accidentes de trabajo?" *El País*, 5 March, 76.

Fundación Encuentro (2000) *Informe España 2000: una interpretación de su realidad social*, Madrid: Fundación Encuentro.

García Delgado, J. L. and Jiménez, J. C. (1999) *Un siglo de España: La economía*, Madrid: Marcial Pons.

García de Polavieja, J. (1998) "The dualisation of unemployment risk, class and insider/outsider patterns in the Spanish labour market," Center for Advanced Study in the Social Sciences *Estudio/Working Paper* 1998/128.

Instituto Nacional de Estadística (INE) (various years, 1971–2000) *Encuesta de Población Activa*, Madrid: INE.

Instituto Nacional de Estadística (INE) (1998) *Contabilidad Nacional de España 1998*, Madrid: INE.

Instituto Nacional de Estadística (INE) (1999) *El Directorio Central de Empresas. Resultados Estadísticos 1999*, Madrid: INE.

MCA–UGT. *See* Federación Estatal de Metal, Construcción y Afines de la Union General de Trabajadores.

Miguélez, F. (1990) "Trabajo y relaciones laborales en la construcción," *Sociología de trabajo* (9): 34–54.

Ministerio de Trabajo y de Asuntos Sociales (MTAS) (1998) *Anuario de Estadísticas Laborales y de Asuntos Sociales*, Madrid: MTAS.

Ministerio de Trabajo y de Asuntos Sociales (MTAS) (1999) *Anuario de Estadísticas Laborales y de Asuntos Sociales*, Madrid: MTAS.

Prieto, C. (1991) "Las prácticas empresariales de gestión de la fuerza del trabajo," in F. Míguélez and C. Prieto (eds), *Las relaciones laborales en España*, Madrid: Siglo XXI, 185–210.

Ruiz, D. and Babiano, J. (1993) *Los trabajadores de la construcción en el Madrid del Siglo XX*, Madrid: Siglo XXI.

Senado de España (1995) *Informe de la ponencia sobre siniestralidad laboral en el sector de la construcción en España*, Madrid: Senado de España.

SEOPAN (1998) *See* Asociación de Empresas Constructoras de Ámbito Nacional.

SEOPAN (1999) *See* Asociación de Empresas Constructoras de Ámbito Nacional.

Union General de Trabajadores (UGT) (2000) *La Temporalidad del Empleo*, unpublished working paper, Madrid: Gabinete Técnico Confederal.

Valdés dal-Ré, F. (ed.) (1993) *Comentarios al convenio general de la construcción*, Madrid: Fundación Anastasio de Gracia.

van der Meer, M. (1998) *Vaklieden en werkzekerheid, kansen en rechten van insiders en outsiders in de arbeidsmarkt*, Amsterdam: Thela Thesis.

van der Meer, M. (2000) "Spain," in B. Ebbinghaus and J. Visser (eds), *Trade Unions in Western Europe since 1945*, London: Macmillan, 573–603.

8 The United States

Dual worlds: the two growth paths in US construction

Peter Philips

INTRODUCTION

The construction industry in the United States solves the challenges of turbulence and uncertainty in two fundamentally different ways. In those regions and sectors of construction where collective bargaining is common, the construction industry carves out career jobs in a casual labor market by attaching well-trained craft workers to craft unions. In those regions and sectors where collective bargaining is rare, the unregulated market fails for the most part to train workers formally, and the industry relies upon a cheaper but less skilled workforce.

The future development of US construction presents a contest between a contractor strategy that relies upon a crew of career craft workers and a contractor strategy that relies upon a handful of key workers backed by a majority of casual and cheaper labor. The presence or absence of government regulation on the wages of construction workers on public works proves to be key to the choice between the high-wage, high-skill path and the low-wage, low-skill path for US construction.

ECONOMIC AND INSTITUTIONAL CONTEXT

In the United States, construction is defined by several characteristics, which together point to the importance of skilled labor within the industry and the challenges faced by the industry in maintaining a skilled workforce. These defining characteristics are turbulence, localness, custom production, and the structure of subcontracting in the construction market.

Turbulence

Construction is the most turbulent and unstable major sector of the economy. This instability is both seasonal – rooted in the effect of climate on outdoor work – and cyclical – the effect of the business cycle and interest rates on the demand for construction.

Construction instability begins with weather. A large proportion of construction – including all road work and much of building construction – is outdoor work exposed to the elements. Because the United States is a large country with distinct weather patterns in different localities, the effect of weather on construction instability varies significantly from north to south and on the coasts vs. the hinterlands. In northern states peak construction employment in the summer tends to be about 30 per cent higher than winter employment, while in states with more moderate weather, such as Florida and California, peak employment runs about 10 per cent greater than off-season employment (Department of Labor, Bureau of Labor Statistics 2001). The annual cycle of taking on and then shedding 10–30 per cent of the labor force is one of the primary employment challenges faced by construction contractors.

Construction also faces unpredictably wide swings in demand associated with the business cycle. For instance, in the highly industrialized state of Michigan, the period between 1978 and 1989 showed a particularly pronounced overall business cycle, tied in part to a crisis in the automobile industry. Overall non-farm employment in the state fell from 3.7 million in December 1978 to 3.1 million in January 1983 and subsequently rose to a new peak of 4 million in November 1989. At the same time, construction fell from 155,000 in September 1979 to 71,000 in February 1983, later rising back to 156,000 in October 1989. Thus, while overall Michigan employment fell by 16 per cent from late 1979 to early 1983, construction employment fell by 54 per cent (Department of Labor, Bureau of Labor Statistics 2001). Climbing back to pre-trough peaks required a 29 per cent increase in overall employment but more than a doubling of construction employment. This is an example of the exponential link between the overall business cycle and the construction business cycle. Compared to overall fluctuations in employment, variations in construction employment show a greater amplitude both seasonally and cyclically.

To respond to these natural and economic gyrations, the construction industry must create a system of deploying capital and labor that is unusually flexible and capable of redeployment quickly across a wide geographic area. As we shall see, such a system entails considerable uncertainty and risk.

Localness

The second defining characteristic of construction is the surprising fact that every locality has a construction industry, usually accounting for approximately 5 per cent of its labor force. In this respect, construction is more like a service sector industry than a goods-producing industry like manufacturing, mining, or agriculture. In the other goods-producing industries, a commodity is made, or a crop is grown, and then it is shipped to the customer. In construction, the work has to go to the customer and the product has to be built there.

Custom production

The third defining characteristic of construction is the fact that most building and road projects are custom-made. Other goods-producing sectors of the economy and agriculture tend to turn out standardized products, but not construction. Every building site and roadbed has unique characteristics requiring customized adjustments. Outside of residential construction, most buildings are individually designed. Architectural and engineering firms exist as a significant auxiliary service industry providing construction with the plans required by these custom characteristics of construction. The non-routine aspects of construction work mean that this industry requires a skilled labor force capable of attending to the changing demands of each custom construction project.

Subcontracting

Finally, the US construction industry is defined by the development of an articulated structure of subcontracting, which developed to meet the challenges of providing a skilled labor force to the construction industry. Owners – those wishing to purchase construction services – hire architects to develop specifications for the desired project. On the basis of these specifications, general contractors submit bids on the project. In the private sector, owners may select a general contractor based on the submitted bid, the reputation of the contractor, past experience with the contractor and possibly other factors. In the public sector, government agencies must select the lowest bidder irrespective of reputation or past experience. This provision is meant to avoid corruption in the allocation of public works to contractors. However, this restriction prevents public owners from selecting contractors based on reputation. Consequently, specifications for projects in the public sector must be highly detailed as a protection against low-bidding contractors recouping profits by reducing the quality of work performed.

The general contractor in both public and private work selects a set of subcontractors to handle specialized aspects of the work. Subcontractors may bid against each other to win the favor of the general contractor, or the general contractor may select subcontractors based on reputation or past business relationships. Both the general contractor and subcontractors may be working on several projects at once. Often contractors find times when they have little or no work at all either due to a slowdown in the local construction market or to the inability of the contractor to win bids or subcontracts.

The need for skilled labor

Thus, in the face of unusual turbulence, the construction industry must make custom-built products with an experienced craft labor force, on-site, in local markets. These requirements put extreme pressure on the construction labor market. Workers must be taken on and shed with the season and the business cycle. Workers must be deployed from place to place as the work moves around. Workers must move from contractor to contractor depending on which contractor is successful bidding on available work. Despite these severe vagaries in work prospects, workers must nonetheless acquire, retain, and develop the needed skills and experience required for high-quality, safe, customized work.

THE CONSTRUCTION LABOR MARKET

There are two fundamentally different ways that the construction industry in the United States attracts and acquires labor in an effort to deploy human capital with the requisite skills to meet the challenges outlined above. Whether contractors acquire workers through the hiring halls of the craft unions or through the more informal methods of the non-union sector, along with the role played by government regulations, determines the extent to which the industry faces "free-rider" problems and the threat of a market failure to train for the future.

Acquiring workers through unions

Most workers come to the contractor once the contractor has won a project, but some key workers stay with the same contractor for many years. In the unionized sector of the construction labor market, which accounts for approximately one-fourth of all construction work in the United States, workers are drawn from union hiring halls, where they have gone once work has dried up on their previous jobs with previous contractors. There are 15 separate unions, one for each major craft in construction. Electricians return to the electricians' union hiring hall; carpenters, to the carpenters' hall, etc. Each locality has a separate hiring hall for each craft. When jobs are scarce, only local members of the union are dispatched from the hiring hall. When jobs are plentiful, the local accepts "travelers" – members of the union from other areas – and these travelers are dispatched from the hiring hall on an equal footing with local members.

Contracts are negotiated locally – in specific cities or regions – between each local craft union and the corresponding contractors who hire workers from that craft. Some unions, such as those for electricians and plumbers, have very decentralized bargaining, while other unions, such as heavy equipment operators (known as operating engineers), elevator constructors, and iron workers, have relatively large jurisdictions for locals, often spanning an entire state or even a multi-state region.

Union specialty contractors are organized generally along craft lines as well, with electrical contractors hiring electricians, mechanical contractors hiring plumbers and/or pipefitters, etc. General contractors tend to hire laborers and carpenters, and they will call to those halls for workers as needed. These contractors form organizations parallel to the respective craft unions for the purposes of collective bargaining as well as providing their members with a variety of industry services.

Both general contractors and specialty subcontractors are free to sign contracts with several unions representing different crafts, giving them the option of hiring a mixed crew of union workers. If a contractor has signed a collectively bargained agreement, the contractor is bound to hire only union members of the craft represented by the agreement in the local area. However, the contractor may hire non-union workers outside the craft jurisdiction of the contract. For example, a general contractor may have both unionized carpenters and non-unionized laborers on the same job. A union contractor is also free to hire non-union members of the craft in other local areas where the contractor is not a signatory to a local contract. Often large regional contractors will be union contractors in strongly unionized areas and non-union contractors in other areas where the union is weak. Furthermore, the general contractor may be a union contractor but some of the subcontractors may be non-union contractors, or vice versa. Thus, specific work sites are often staffed by a mixture of union and non-union workers.

Contractors become union contractors in one of two ways. Either the contractor's workers have voted in a National Labor Relations Board (NLRB) election to be represented by a union, or the contractor has simply signed up with a specific union accepting the collectively bargained agreement previously negotiated between the union and the local contractor organization. In reality, NLRB elections are rare in construction due to the short duration of specific building projects and the rapid turnover of workers. By the time an election can be held, the project may well be over and the workers on their way to other work with other contractors. Therefore, the primary source of unionization in construction is the willingness of contractors to sign up with unions.

Contractors are willing to sign up with unions because they provide a reliable stream of qualified workers to staff work as it comes on line. Each craft union sends out to contractors only workers the union has qualified, either through an apprenticeship program or through some set of tests that qualify the worker for journeyworker status. The union will also send out apprentices enrolled in a joint labor–management-operated apprenticeship program.

While approximately 25 per cent of the US construction labor force is unionized, union membership varies widely by state, with over 50 per cent in Illinois and less than 2 per cent in North Carolina. It also varies widely by craft, with trades found primarily in the industrial sector, such as iron workers, having a 70 per cent unionization rate while crafts found primarily in residential work, such as carpet layers, have a unionization rate of less than 10 per cent.

Acquiring workers in the non-union sector

In the non-union sector, there is no formal system of labor acquisition. Some very large non-union contractors attempt to hold on to workers throughout the year in a manner similar to that of employers outside construction. Their strategy is to try to stay busy across the year and across the business cycle by working in many states with differing weather patterns and differing business conditions.

They cross-train their workers to gain greater flexibility in their workforce. For example, a worker may begin working a job as a heavy earth-moving equipment handler and then move to exterior structural steel work and then to interior carpentry. Thus, in contrast to the traditional hiring hall approach, which has adapted to the turbulence of construction, some non-union contractors attempt to overcome the turbulence of the construction industry by developing permanent attachments between worker and employer.

At the other end of the spectrum, some very small and local non-union contractors also attempt to establish permanent employer–worker ties by meeting only the most stable of local construction demand – typically service, maintenance, and repair work. These contractors shun the big jobs found in the boom in order to avoid hiring workers whom they cannot retain beyond the peak demand periods.

While some very large and very small contractors succeed to some extent in avoiding the turbulence of construction and establishing relatively permanent employer–worker ties, the majority of contractors cannot do this. Furthermore, if they tried, the unstable portion of construction demand – a very large percentage of all demand for construction services – would go unmet.

For the majority of non-union contractors, therefore, the periodic taking on and shedding of workers in tune with the demand for construction is one of the essential business dilemmas they face. Occasionally, temporary labor agencies attempt to provide hiring-hall-like services to non-union contractors, but they are often not in a position to vouchsafe the credentials of these workers. Sometimes, extended family and ethnic minority networks serve as informal institutions providing workers on an as-needed basis in some areas. Generally, however, in the non-union sector, workers come with unclear credentials in a haphazard fashion, and, as we shall see, non-union contractors are reluctant to train workers who will leave them at the end of the current project.

Labor market regulations

There are a limited number of regulations governing the construction labor market in the United States. The contractor must have a license and in most cases must secure bonding that insures against defaulting on work. In many, but not all, states, electricians and plumbers also must be licensed by the state or in some cases by the municipality. However, only rarely are other crafts required to undertake training and/or pass tests to obtain craft licenses. Thus, the state does not take on the role of certifying the credentials of most construction workers. In the union sector of the industry, union membership involves meeting union-set criteria for qualifications. In the non-union sector, contractors must rely upon references from other, often competing contractors and other informal tests of a worker's qualifications.

In the public sector in about 60 per cent of the states, wage rate payments on public construction are set by prevailing wage regulations. These regulations require the contractor not to undercut local wage rates and in many cases to pay wages and benefits agreed to in local collectively bargained contracts, regardless of whether or not the contractor is a signatory to that contract. In the private sector, union contractors must pay the local collectively bargained wage rate, but non-union contractors are free to offer any wage above the federal minimum wage.

Thus, relative to other countries, the United States has an unregulated construction labor market. No systematic government-sponsored training is provided to the industry, although in some localities technical schools and community colleges provide some training programs, partially financed by local public funds and partially by student tuition.

Unemployment and turnover

One result of the turbulence in the construction industry is unusually high levels of unemployment and labor turnover for construction workers compared to other workers. In 1994, the overall unemployment rate for men was 5.4 per cent, but for men working in construction occupations it doubled to 11.1 per cent (Bureau of the Census 2001).[1] Similarly, in 1999, when the overall unemployment rate had fallen to 3.7 per cent for all male, non-construction workers, it was 6.8 per cent for male construction workers.

Male construction workers who were unemployed were roughly three times more likely to be unemployed because of job layoff compared to other males. In 1994, only 11 per cent of non-construction unemployed men were unemployed because of job layoff, while fully one-third of all male construction workers were unemployed because of layoff. In 1999, even though overall and construction unemployment rates were lower, a similar pattern emerged. Only 15 per cent of all unemployed men had been laid off, while 40 per cent of all male construction workers were unemployed because of layoffs. Thus, construction workers at any given time are twice as likely to be unemployed and roughly three times as likely to be unemployed because of layoffs.

High labor turnover and high unemployment rates are problems not only for the specific workers involved but also for the construction industry itself. As a craft industry, construction relies upon the skills and knowledge of the labor force to insure efficient, safe production and quality output. Unemployment is like an economic earthquake that rocks the foundation of these skills. Laid-off workers lose valuable experience. Unemployed yet skilled workers are tempted to take their skills elsewhere. It is difficult to coax talented people into making a career commitment to construction work when that work is highly unstable.

THE CHALLENGE OF HUMAN CAPITAL ACCUMULATION

Against this economic context, the construction industry in the United States has grappled with the problem of how to accumulate human capital. Human capital is a form of productive capital just as machinery and equipment are. It consists of skills, knowledge, experience, and judgment acquired by and residing in human beings. Human capital is a crucial ingredient needed to make physical capital work. In construction, at least, state-of-the-art tools and equipment do not run themselves. They require skilled craft workers to make them effective implements of production. Trained, knowledgeable, and experienced construction workers bring out the best in the materials bought for the construction process. Skilled, professional construction workers play a major role in seeing to it that all of the construction ingredients come together properly.

Much in construction militates against the accumulation of human capital. This industry runs at double the unemployment rate of the overall economy. It generates three times the amount of involuntary part-time work of the overall economy. It usually entails casual, temporary relationships between workers and contractors. It is directed by contractors who on average employ seven or eight workers in building construction and 20 workers on highway construction. It tolerates a high rate of firm turnover; many contractors in business today are not those in business 10 years ago. It tolerates a high rate of labor turnover; the worker trained today easily can become a competitor's worker tomorrow. These factors are the ingredients for a casual labor market where employers are not committed to workers and workers are not committed to the industry. Lack of commitment leads to lack of investment in human capital specific to the needs of the construction industry on the part of either the contractor or the worker.

The construction industry sells big-ticket products that last a long time. So customers today will not often directly benefit from the training of the next generation of construction workers. It sells products where the effect of shoddy craftsmanship may not show up immediately, may be easy to hide, and may not affect the first owner of the construction project. Finally, the product is often sold in a low-bid process that favors contractors who exclude downstream costs from the cost of today's building. These factors militate against the owners of construction projects stepping in to see that the long-term training costs of the industry are paid for in the short-run context of the owner's project.

So who is going to pay for the training that is needed? The problem is how economic incentives can be put into place that will induce someone – the worker, the contractor, or the owner – to pay for this needed human capital accumulation.

The free rider problem: human capital in the open shop

Left to its own devices, construction in the United States tends to slide down a low-skill, low-productivity, low-quality, and low-wage development path. The root of the problem lies in an endemic construction market failure associated with training and the resulting "free-rider" strategies of all participants, all of whom seek a "free ride" from paying the long-term costs of training in construction.

The individual worker is discouraged by high rates of unemployment and underemployment and frequent worker movement out of the industry. It is difficult to get young people to invest in themselves for a future in such an industry. The human capital accumulated by the worker's effort and expenditure might lay idle and unrewarded. Idle capital means lost returns. Individual workers are reluctant to invest time and money in an endeavor that promises high rates of lost returns.

The owner is stuck with yesterday's accumulation of human capital in construction, whatever that might be. Any money the owner contributes to training will benefit future buyers of construction but not the owner today.

Under the pressure to win bids in the short run, contractors hesitate to embed in their bids the long-run costs the industry must face if it is to train the next generation of construction workers, keep the current labor force safe and healthy, and sustain the previous generation of construction workers in their retirement years. They tend to compete over who can best jettison the long-term needs of the industry to meet the short-term goal of winning contracts.

Bids formulated with only the short run in mind steadily undercut the foundations for a strong and safe construction industry. The market fails to train adequately as contractors strive to take free rides on the training provided by their competitors. The Business Roundtable, a group of large industrial and commercial consumers of construction services, described the state of the free-rider problem in the 1990s:

> Companies are currently experiencing significant problems in staffing construction projects, resulting in escalating costs and costly schedule delays... In late 1996, The Business Roundtable surveyed its member companies... Over 60 percent of the survey respondents indicated they had encountered a shortage of skilled craft workers, and 75 percent reported the trend had increased over the past five years... The union sector has always excelled in craft training through the joint labor/management apprenticeship programs... the open shop, as a whole, has not supported formal craft training to the extent necessary. They have succeeded by attracting skilled workers from the union sector as market share shifted and recruiting skilled workers from competitors as individual workload changed. As the well begins to dry up, the ability to use these methods decreases... Through the years, the subject of funding for training has come up repeatedly. All of the discussion has been on the open shop side. Training on the union side has always been required and paid for by the owner. A trained workforce was expected and guaranteed by the contractors with costs passed on to the owner as part of the collective bargaining labor rate. It has been a different story on the open shop side (Business Roundtable 1997: 2–14).

Shared costs: human capital under collective bargaining

Collective bargaining cuts through this Gordian knot of free-rider problems and overcomes the failures of the market to provide adequate training for the construction labor force. Through collective bargaining, the long-term costs of the industry – the costs of training, health insurance, and pensions – are embedded in every short-term bid made by contractors who are signatories to these contracts.

In the US construction workers organize into an occupational or craft union, and contractors organize into a multi-employer association along craft lines for specialty contractors and across craft lines for general and heavy-and-highway contractors.[2] These collections of individuals and companies then engage in local area bargaining, which results in contracts that typically span 1–3 years. These local negotiations, occupation by occupation, result in hundreds of distinct agreements across the country.

One of the most common and widespread results is an agreement between contractors and workers to establish a joint labor–management-operated, multi-employer apprenticeship program. Under these contracts, each contractor who is signatory to the collectively bargained agreement contributes a contractually agreed-upon and specified amount of money, usually ranging from 25 cents to $1.00, for each hour of work done by each worker employed by that contractor. Apprentices are rotated among employers to expose apprentices to a wide range of work situations faced by a given craft, and the employer group shares the skilled workers generated by this training system.[3] Contractors also agree to include a fixed amount per hour worked for the maintenance of a health insurance program and another fixed amount to cover the cost of pensions for retired workers.

The collectively bargained contract resolves the free-rider problems regarding training in construction. All the employers signing the contract know that the long-term costs they put into their bids will also be in the bids of competitors. Therefore, each contractor can prepare a bid for a new project confident that competitors will not win the project based on skimping on training.

Indeed, the incentives are redrawn under this free-market contract. Contractors who sign these agreements compete with each other over who can staff the job with the fewest labor hours. They are motivated to equip workers with the best tools, equipment, and materials, and to obtain the best-trained workers so the tools and equipment provided are well handled. By equipping workers with the latest technology, the contractor continually exposes them to innovations that expand their skills and experience. Consequently, the collectively accumulated pile of human capital grows through the quality of experience as well as the quality of training.

This creates a mutually reinforcing system where the contract insures that employers collectively contribute to human capital investment, and the incentives within the contract encourage the best use of the labor that is trained. This pushes technological change and sets construction upon a high-skill growth path. Over the long haul, this system of competition encourages skill creation and retention – the basis for a well-paid local construction workforce.

Because these systems are established through bargaining with a union, it is not quite clear who is paying for this human capital accumulation – the contractors or the workers. However, what is abundantly clear is the fact that these hourly contributions form a scholarship fund or subsidy to young persons. By investing as a group in the training of young people in the industry, employers are making a credible promise of future employment. While any one contractor can go out of business, the collection of employers has a more stable future and a more reliable demand for labor. Thus, multi-employer apprenticeship programs tease out of young people a greater willingness to invest in themselves for a future in construction.

HUMAN CAPITAL ACCUMULATION UNDER COLLECTIVE BARGAINING

In states where collective bargaining is common in construction, we find a chain of favorable consequences. With more extensive participation in apprenticeship programs, higher education levels, and a substantially higher investment of capital per worker, the construction industry under collective bargaining fosters a workforce that can better utilize new equipment and technology. This workforce, not surprisingly, is substantially more productive.

Apprenticeship program participation

Figure 8.1 shows the number of newly enrolled construction apprentices in the United States for each year since 1989. These data represent 70 per cent of all construction apprenticeship programs in the United States and are broken down by the apprentices employed by union and non-union contractors. While the

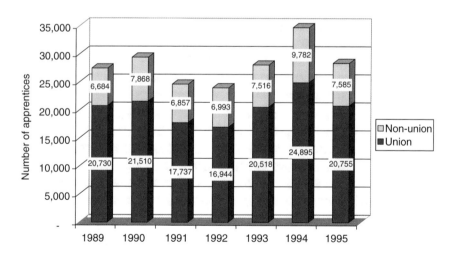

Figure 8.1 Annual new enrollments in construction apprentices.

Source: Department of Labor, Bureau of Apprenticeship Training 1995. These data represent approximately 70 per cent of all construction apprenticeship programs in the United States and exclude data from California and New York. Similar results are found using Kentucky state apprenticeship data in the Londrigan and Wise study (1997).

number of new apprentices entering construction varies with the construction business cycle, the proportion trained under collective bargaining remains roughly the same. Approximately three out of every four new apprentices enroll in programs created by collective bargaining.

A more important question, however, is not enrollment but the graduation of apprentices. As Figure 8.2 indicates, collectively bargained programs turn out 82 per cent of all construction craft workers trained through apprenticeship in the construction industry. In some crafts, open shop apprenticeship programs account for only 1–2 per cent of all the apprentices graduating to journeyworker status. For instance, only 1–2 per cent of the apprentices graduating to journeyworker status among operating engineers or structural steel workers comes from open shop apprentice programs. Only 9 per cent of the graduating bricklayers and 8 per cent of the graduating carpenters come from open shop apprenticeship programs. Even among plumbers, where the open shop has its largest share of graduating apprentices, two-thirds of all plumber apprentices graduate from collectively bargained programs.

Investment in formal education

The specifications of the collectively bargained contract compel contractors to invest in young workers. This willingness, in turn, attracts young workers who

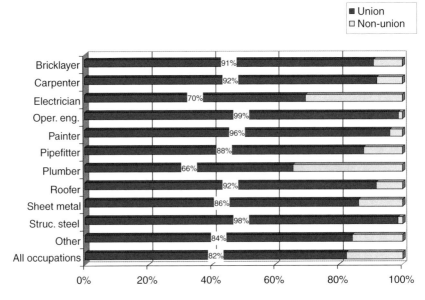

Figure 8.2 Relative contributions of collectively bargained and open shop programs to graduating journeyworkers at the end of 1995 for classes entering in 1989, 1990, and 1991.

Source: Department of Labor, Bureau of Apprenticeship Training 1995.

are willing to invest in themselves. This is reflected in the difference in formal education between workers under collective bargaining compared to workers in the open shop sector of construction. Figure 8.3 shows that among all workers under collective bargaining, 13 per cent have not obtained a high school education. Among open shop workers, more than twice as many, 28 per cent, have failed to graduate from high school. At the other end of the spectrum, while only one out of every four open shop workers has at least some college-level education, the rate increases to one out of every three workers under collective bargaining.

Construction apprenticeship training is the largest system of privately financed higher education in the United States. Training lasts from 3 to 5 years depending on occupation. Entering classes run about 40,000 students per year. At any one time, roughly a half million apprentices are enrolled. Collective bargaining accounts for 75–80 per cent of these enrollments. These multi-employer programs have a higher graduation rate in part because they have higher admission standards. This is part of the meaning of the data in Figure 8.3. Also, jointly managed labor–management programs have higher graduation rates because they offer scholarships that show a higher educational commitment toward their students. This encourages young persons who have attained formal educational degrees to invest that human capital in the construction industry. By investing in the young, contractors who use collective bargaining get the young to invest in themselves.

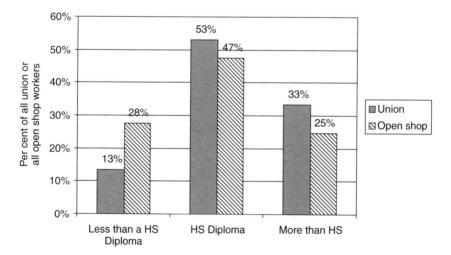

Figure 8.3 Distribution of educational attainment of workers under collective bargaining and in the open shop, 1994–9.

Source: Bureau of the Census, Microdata Access Branch (2001).

Social insurance provision

In the United States, health insurance, if it is provided at all, is typically provided through the employer. But with rapid labor turnover in construction, many non-union contractors find it difficult to provide health insurance for any but the key workers who stay with the contractor throughout the year.

Figure 8.4 shows how collective bargaining overcomes the market failure to pay health insurance and pension benefits in construction. The x axis in the figure shows the percentage of construction workers who work under collective bargaining in each of the 50 states and the District of Columbia (Bureau of the Census 2001), calculated by averaging data for the period 1994–9. The y axis shows voluntary benefits paid by the employer as a percentage of worker payroll – i.e., wages and salaries in 1997 (Bureau of the Census 1997). Voluntary benefits consist primarily of health insurance, pension contributions, and apprenticeship training contributions. These benefit payments are distinguishable from legally required benefits – Social Security contributions, workers' compensation insurance for injuries, and unemployment insurance. Voluntary benefits paid by the employer clearly rise in relative terms with increased collective bargaining.

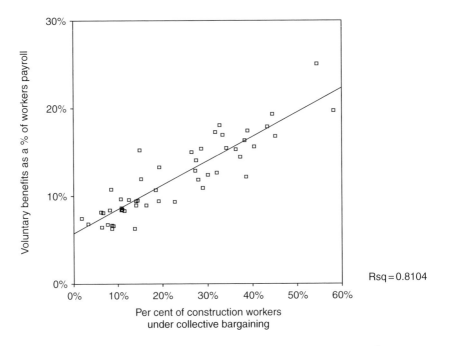

Figure 8.4 Voluntary benefits (health insurance, pension benefits, apprenticeship training contributions, etc.) as a percentage of take-home pay by percentage of construction workers in a state who are covered by collective bargaining.

Source: Bureau of the Census 1997.

In addition, the actual dollar value of benefits also rises with the greater percentage of construction workers in a state working under collective bargaining. For instance, in states with very low rates of collective bargaining, such as North Carolina and South Carolina, contractors pay approximately $1500 on average for health insurance, pension contributions, and training. In states with high rates of collective bargaining, such as Illinois, contractors pay around $5000–6000 per worker for these same benefits (Bureau of the Census 1997).

Marital status

The payment of these family-friendly benefits allows young people who enter the US construction industry to remain after they marry and form a family. Figure 8.5 shows that workers under collective bargaining tend to be married, while their counterparts in the open shop are less likely to be married. One-third of open shop workers have never married, compared to only one-fifth under collective bargaining. Fifty-three per cent of the open shop workforce is currently married, compared to 67 per cent under collective bargaining. The collectively bargained labor system permits and facilitates greater family formation and allows workers to stay within the industry as they age and gain experience within the industry.

Age distribution

Figure 8.6 demonstrates that the age profiles of union and non-union construction workers are significantly different. The open shop labor force consists of twice as many teenagers compared to the union workforce. In general, the open

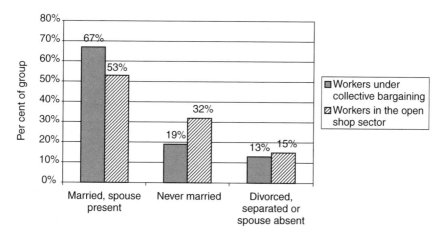

Figure 8.5 Comparison of marital status under collective bargaining and in the open shop sector.

Source: Bureau of the Census, *Current Population Survey*, Outgoing Rotational Group, 1994–9.

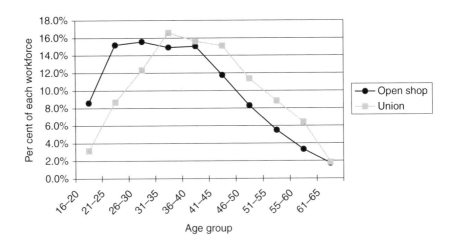

Figure 8.6 Age distribution of union and open shop construction workforces, 1999.
Source: Bureau of the Census, Microdata Access Branch 2001.

shop workforce uses more under-30 workers and fewer over-30 workers. Thus, the non-union construction labor force is younger, less married, and less formally trained, and it receives less apprenticeship training.

THE RESULTS OF HUMAN CAPITAL ACCUMULATION: GREATER LABOR PRODUCTIVITY

Variations in the prevalence of collective bargaining across states can be used to estimate the effect of these types of market contracts on construction productivity. Figure 8.7 shows statewide averages in the percentage of union members among all construction workers. Unionization rates vary from 2 per cent in North Carolina to 58 per cent in Illinois. These are proxy measures of the prevalence of collective bargaining in each state's construction labor market.

Table 8.1 shows the results of a set of simple least squares linear regression lines relating the extent of collective bargaining in each state to various measures of capital investment per worker or labor productivity. The extent of collective bargaining is measured by the percentage unionized of all construction workers in each state and comes from averaging results over 6 years (1994–9) of the *Current Population Survey* by the Outgoing Rotational Group. The investment and output measures are from the 1997 *Census of Construction, Geographic Series*. The results are reported separately for heavy-and-highway contractors, specialty contractors, and general contractors. Table 8.1 shows numerically the results of statistically drawing a line (a "linear regression") through a scattering of points where, in each case, the extent of collective bargaining is measured on

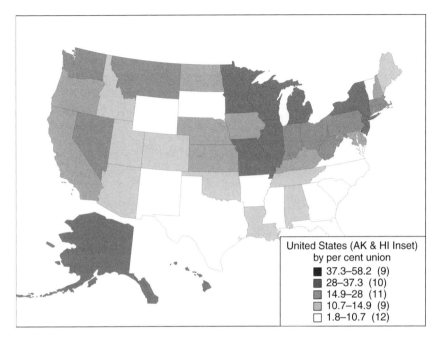

Figure 8.7 Prevalence of collective bargaining by state as shown by the percentage of union members of all construction workers in each state.

Source: Bureau of the Census, Microdata Access Branch 2001.

Note: There is considerable variation across the United States. Illinois has the highest unionization rate, at 58 per cent of all construction workers. North Carolina has the lowest prevalence of collective bargaining in construction, with a unionization rate of 2 per cent. Unionization rates are typically higher in all nonresidential segments of construction simply because collective bargaining is uncommon in residential construction. When this roughly 20 per cent of construction is eliminated from consideration, the estimate of collective bargaining coverage rises in each state.

the *x* axis and then, in order, the *y* axis measures (1) the value of capital stock per worker; (2) the value of annual capital expenditures per worker; (3) the value of output per worker; (4) the value added per worker; and (5) the value of materials used per worker.[4] All of these values are in 1997 dollars. The slopes of the lines drawn are reported in the second column, labeled "elasticity."

This column of slopes is called elasticity because in all cases in Table 8.1, rather than taking the actual per cent unionized on the *x* axis and taking the actual values of various measures on the *y* axis, the natural logs of these measures are taken. Using this technical trick means that the slopes of the lines drawn are "elasticities." An elasticity is a number that shows the percentage change in something due to a percentage change in something else. In our first example in column 2, row 1, the number 17 per cent shows the percentage change in the

Table 8.1 The effect of collective bargaining on the capital stock per worker, annual
capital expenditures per worker, the value of output per worker, value added
per worker, and the value of materials installed per worker

Effect of an increase in CB on	Elasticity(%)
Heavy-and-highway contractors	
Capital per worker	17
Capital expenditures per worker	16
Output per worker	18
Value added per worker	20
Value of materials per worker	13
Specialty contractors	
Capital per worker	15
Capital expenditures per worker	8
Output per worker	14
Value added per worker	17
Value of materials per worker	11
General contractors	
Capital per worker	9
Capital expenditures per worker	7
Output per worker	4
Value added per worker	9
Value of materials per worker	−3

capital stock due to a doubling or 100 per cent increase in the per cent unionized. So, for example, if one doubles the unionization rate, say from 10 per cent
to 20 per cent of all construction workers, in the case of heavy-and-highway
contractors, the capital stock invested per worker will rise by 17 per cent.

In all cases except the last three, results of standard tests indicate that the
elasticities estimated in column 2 are statistically significant.[5] For instance, for
heavy-and-highway contractors, if you double the extent of collective bargaining, the resulting increased wages and skills of the transformed labor force will
induce heavy-and-highway contractors to increase the capital stock per worker
by 17 per cent and induce specialty contractors to increase the capital stock by
15 per cent. Capital expenditures per year will increase by 16 per cent for heavy
contractors and 8 per cent for specialty contractors. In the case of general
contractors, a doubling of the extent of collective bargaining will result in
a 9 per cent increase in the capital stock and a 7 per cent increase in capital
expenditures per worker.

Why do we find strong connections between the use of collective bargaining
and the amount of capital invested per worker among heavy-and-highway
contractors and specialty contractors, but only weak connections among general
contractors? The answer lies in the fact that subcontracting is evolving. In building construction, general contractors are off-loading actual construction work to
subcontractors. The value that they bring to the table is increasingly that of
expert buyers of materials and sellers of subcontracts rather than as producers of

actual construction work. Therefore, their capital stock purchases as a group are less tied to collectively bargained contracts regulating actual construction activity.[6] For the general contractor who continues to perform actual construction work rather than job it out, the relationship between collective bargaining and investment per work may indeed hold. However, in the *Census of Construction*, self-performing general contractors cannot be isolated from those who manage the job but subcontract out the actual work.

In general, we find that a doubling of the practice of collective bargaining leads to a 10–20 per cent increase in labor productivity among the workers of specialty and heavy-and-highway contractors. This may also be true of self-performing general contractors. If so, however, the effect cannot be seen in data that includes general contractors that outsource actual work.[7]

PUBLIC POLICY TOWARDS THE CONSTRUCTION INDUSTRY

This system for overcoming market failures in accumulating human capital within the construction industry depends upon prevailing wage regulations as the single most important government policy promoting the practice of collective bargaining in construction. As the tide has turned against the use of prevailing wage laws in many states in the United States, we can see how policy makers in these states have placed themselves in a dilemma regarding how to ensure that the long-term needs of the construction industry are met.

The effect of prevailing wage regulations

Prevailing wage regulations, which require contractors on public works to pay the local prevailing wage for each craft on the construction site, comprise the primary public policy supporting collective bargaining in US construction. The prevailing wage rate is often, but not always, the collectively bargained wage rate. Where union density is sparse, often the prevailing wage rate is the average wage rate for a particular construction occupation. However, where the prevailing wage rate is the collectively bargained rate, the construction industry itself is usually required to embed in bids the long-term training, health, and pension costs of construction. Non-union contractors are forced to follow suit on public projects governed by prevailing wage regulations and sometimes adopt the high-skill strategies of union contractors and follow suit on private works as well.

However, where prevailing wage laws were never enacted or have been repealed, collective bargaining is uncommon. Absent these contracts, worker health insurance is uncommon or focused on only a minority of key workers. Retirement pensions are also uncommon, putting pressure on government agencies to underwrite the care of retired workers.

The federal government has implemented a prevailing wage law, the Davis-Bacon Act, since 1931, and most states passed prevailing wage regulations between 1891 and 1970. However, of the 50 states in the United States, 9 states never passed prevailing wage regulations. Another ten have eliminated their regulations in the last two decades.

From this variation in regulations across states, we can measure the effect of this policy on construction apprenticeship training. The Department of Labor's Bureau of Apprenticeship Training monitors registered apprenticeship programs – both union and non-union – in the construction industry. Data are available for 1975–8 and 1987–90. Not all states have reported to the Bureau of Apprenticeship Training for all years during these periods. Nonetheless, 29 states did report registered construction apprentices for every one of those years. These states included six states that eventually repealed their prevailing wage laws, four states that never had prevailing wage laws, and 19 states that retained a prevailing wage law throughout the period. These 29 states can be divided into the categories "repeal," "never-had," and "retained-law," for comparison. No state had repealed its prevailing wage law by 1978. By the end of the first quarter of 1987, all nine repeal states had passed their repeals except Louisiana, which repealed in 1988. The data for 1987 are for the summer of 1987, after Kansas had repealed in that year.

Figure 8.8 shows that states that repealed their law saw their apprenticeship training rates fall dramatically. Before repeal, these states had apprenticeship training rates similar to other states with prevailing wage laws. After repeal, these states had training rates similar to states without prevailing wage laws.

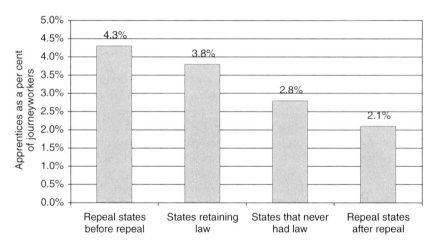

Figure 8.8 Apprentice training rates as a percentage of journeyworkers by legal climate related to prevailing wage law.

Note: Training rates were determined by taking the number of enrolled apprentices in a state according to Department of Labor's Bureau of Apprenticeship Training (1995) data and dividing by the number of construction workers in that state according to Bureau of Labor Statistics (2001) data from the *Current Employment Statistics.*

In general, states with prevailing wage laws trained around one-third more apprentices than states without prevailing wage laws.

The mechanism that generated this loss of training was direct and indirect. In some cases, state laws required or allowed, as part of the prevailing wage regulation, the payment of apprenticeship training contributions. Such contributions fell off among non-union contractors working public jobs after the repeal of these states' prevailing wage laws. Furthermore, to the extent that the repeal of prevailing wage regulations shifted the weight of construction work away from collective bargaining, these laws fostered market failure – a failure to train associated with free-rider problems within the open shop sector of construction.

Skills shortages

The problems associated with loss of training due to legal policies adopted in the 1980s came home to roost in the 1990s, when the US construction industry expanded. For instance, *Engineering News Record*, a major construction industry publication, reported in 1995:

> The industry has known for much of the past decade that it was headed for manpower trouble when the business cycle turned up. That has come to pass. Contractors in some areas now are knee-deep in labor shortages, according to an ENR survey of the Top 400 US general contractors and the Top 600 specialty contractors . . . Nonunion contractors working in bustling areas appear to have the biggest manpower problems, according to the survey results. For example, 56% of the union crafts in the West reported having no labor shortages while only 10% of the open shop crafts have no problem. Only 10% of the union crafts have a severe craft shortage problem there, while 29% of the nonunion crafts are severely short . . . The spreading craft labor shortage problem is underscored by the results of an open shop survey conducted by the National Center for Construction Education and Research, Rosslyn, Va. Of 2,437 responses, 1,808 or 74.2% reported shortages in their area for 14 crafts (Krizan 1995).[8]

Erosion of wages

This craft shortage and dearth of training in the mid-1990s came in the face of a long-term decline in construction wage rates and incomes compared to other prospective wage rates and incomes in the US economy. In May 2000, *Business Week* reported:

> It's a puzzle. New homes sales and housing starts have been on the high road for years, construction trades unemployment has hit its lowest level in at least two decades, and home builders grouse that labor shortages have added 20 days to the time needed to build a single-family home – boosting costs.

Yet as Elliot Eisenberg of the National Association of Home Builders notes in a new report, the relative pay of construction workers has been falling for two decades, from 20% more than the median weekly wage of all workers in 1983 to just 3% more in 1999, even as shortages have become acute. Moreover, the decline has occurred among highly paid plumbers and electricians, as well as among lower paid painters and dry-wall installers.

What's not clear is the cause of the long-term erosion in hard-hat pay. Although some think that a shift toward more construction in less-unionized states is a factor, Eisenberg suggests that the slippage in relative wages may reflect a drop in the skill levels of workers entering the construction field (Koretz 2000).

Thus, public policies discouraging collective bargaining have had a double role in helping create the current labor shortage in US construction. First, the absence of collectively bargained contracts fostered a market failure to train. Second, the absence of collective bargaining has eroded the wage rate advantage construction has historically had over competing industries. With this advantage diminished, other competing industries look more favorable because they are steadier and safer, and they provide better benefits.

A primary effect of these lower wages relative to other sectors of the economy is not only that construction is losing skilled workers to other industries but also during boom periods such as today, construction is not able to attract skilled and experienced workers from other industries. The construction expansion is being staffed by young, inexperienced, and less trained workers. Jeffrey M. Robinson of Personnel Administration Services, Inc. (PAS) in Saline, Michigan, surveys open shop contractors. In 1998, *Engineering News Record* interviewed him:

The open shop may be booming, but it is not measuring up in manpower. Contractors faced with craft labor shortages are boosting wages to keep workers from jumping ship and to attract new ones . . . The trends may be somewhat subtle, but these are "pretty exciting" times for craft compensation, says Jeffrey M. Robinson . . . The "tip off" is that the average 1998 craft pay hike [among open shop contractors] is expected to exceed that for executives for the first time in the 15 years that PAS has been surveying nonunion craft wages . . . In general, the PAS survey indicates that contractors "are paying whatever it takes to get people" and construction slowly is closing the pay gap with other industries, says Robinson. That process of normalizing construction wages should take another three to four years at the current rate of escalation, he says. The industry is not able to steal workers away from other industries but the survey shows "an influx of a lot of new employees this year," says Robinson (Krizan 1998).

Two years later, in June 2000, after conducting another survey of open shop wages, Mr. Robinson was again interviewed:

> Sooner or later, everyone is going to have to address the issue of base pay... When employees have the opportunity to go to a more stable work environment in industry or manufacturing, why shouldn't they? (Krizan 2000)

As an alternative to turning to younger workers or higher wages, open shop contractors have turned to foreign workers, especially in southwestern states. For example, in December 1998, *Texas Construction* reported:

> With record low unemployment rates in Texas, an Austin firm is offering labor-needy construction companies with a possible solution. Foley Enterprises is expanding its work permit services to the construction industry. Foley is able to acquire legal work visas for companies feeling the labor crunch ("Austin Firm Addresses Labor Shortages," *Texas Construction* 1998).

Figure 8.9 shows that union and non-union contractors employ almost the same percentage of racial minorities. Among the construction workers hired by open shop contractors, 7.9 per cent were racial minorities – Blacks, Asians, and

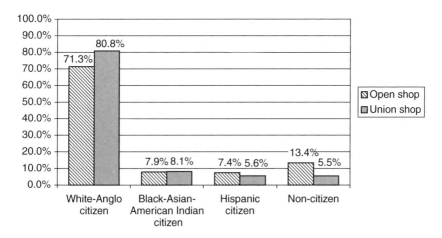

Figure 8.9 Comparison of the racial and ethnic composition of union vs. non-union workforces, 1999.

Source: Bureau of the Census, Microdata Access Branch 2001.

Note: While racial minorities (Blacks, Asians and American Indians) account for the same percentage of both the open shop and union workforces (i.e., around 8 per cent nationally), there are more White citizens among union workers compared to open shop workers (81 per cent compared to 71 per cent). Most of this differential is accounted for by the fact that the open shop workforce employs relatively more non-citizens (13.4 per cent compared to 5.5 per cent for union shop contractors).

American Indians – and US citizens. A similar percentage, 8.1 per cent, of all union workers were racial minorities. However, there were substantially more white, non-Hispanic workers who were citizens working under collective bargaining compared to white, non-Hispanic citizens working in the open shop (81 per cent compared to 71 per cent). This differential is accounted for primarily because open shop contractors hired substantially more non-citizens (13.4 per cent compared to 5.5 per cent).

Public policy decisions

Like the provinces in Canada, but unlike many other more politically unified countries, the regulatory environment of the construction industry in the United States varies considerably across states. Looked upon closely, the US construction industry is, in fact, composed of many industries. They vary by market type – residential construction, small commercial building projects, large commercial building projects, the construction of industrial facilities, and the construction of roads, dams, and other infrastructure. These many industries that make up construction also vary by state, based on the presence or absence of prevailing wage regulations and a corresponding presence or absence of collective bargaining.

Policy makers in both states with and states without prevailing wage regulations continually face the question whether to regulate or to deregulate construction. Advocates of prevailing wage law repeals assure elected officials that substantial cost savings can accrue to jurisdictions that eliminate these regulations. These promises of cost savings assume that the dramatic cuts in wages, benefits, and training contributions associated with deregulation will not substantially affect productivity. Further, these claims of cost savings presume that the government will not be asked to fill the void created by the destruction of market-based training and insurance. In short, the claims of repeal advocates ignore the consequences of unregulated free-riding among construction contractors.[9] In effect, repeal advocates encourage the industry to jettison long-term training costs with no clear prescription as to who picks up the tab.

CONCLUSION: PAY ME NOW OR PAY ME LATER

The US construction industry is becoming a two-tiered industry. In regions and sectors where collective bargaining is common, construction is developing along a high-wage, high-skill, capital-intensive, and human-capital-intensive path. In regions and sectors where the open shop predominates, construction is developing along a low-wage, low-skill, labor-intensive path. As a consequence, along the high-skill path under collective bargaining, the following holds true:

- Workers are one-third more likely to have some college education.
- Workers have substantially more apprenticeship training.

- Contractors invest from 9 per cent to 17 per cent more capital per worker.
- In the case of heavy-and-highway contractors and specialty contractors:
  - 11–13 per cent more materials are put in place by each worker.
  - Contractors get 14–18 per cent more output from each worker.
  - Contractors get 17–20 per cent more value added per worker.

In contrast, in the open shop, the following holds true:

- Workers are twice as likely to be high school dropouts.
- Workers are twice as likely to be young and inexperienced.
- Workers are two-and-one-half times as likely to be a non-citizen.

The US case presents a model where collective bargaining rather than government programs provides the institutions and arrangements required to embed the long-term training, health, and pension costs of construction into the short-term calculations of contractors and owners. Where collective bargaining reigns, the true costs of construction are paid up front. However, collective bargaining is on the decline – in part associated with the erosion of government support of collective bargaining generally and specifically in the repeals of prevailing wage regulations in various states. Where the open shop prevails, the long-term costs of maintaining the health and skills of the labor force are put off or never paid at all. Therefore, the US construction industry faces a classic case of "pay me now or pay me later," with many in the open shop sector of construction hoping that "later" never comes. Increasingly, however, the purchasers of construction services are concerned about the quality of the labor force that produces their buildings and roads. They worry about the downstream costs of maintenance and productivity costs of infrastructure with doubtful quality built into it. The abundant and destructive free-rider strategies found in construction need to be brought under control through appropriate regulations. While direct regulation of training, health, and pension arrangements is not necessarily required for a healthy construction industry, background regulations promoting collective bargaining are a needed buttress to make the US model work well.

NOTES

1 Male workers only. The category of construction workers excludes non-construction-site workers such as office workers, architects, and engineers.
2 Outside the North American context, heavy-and-highway contractors are often referred to as "civil engineering contractors."
3 Typically, when non-union contractors operate apprenticeship programs, those are single-employer programs – i.e., programs where apprentices are limited to the business activity of a single contractor.
4 The capital stock is the book value of depreciable assets excluding land at the end of 1997. Capital expenditures are the purchase of capital stock excluding land during the year. Value of output nets out the value of output subcontracted to others. Value added

is the value of output minus the purchase of construction materials and supplies. It includes wages, capital depreciation, and contractor profits. The value of materials includes construction materials and supplies. Unionization rates are calculated for construction workers only and exclude other employees. "N" refers to the number of states reporting data in each regression. Missing states are due to missing data in the *Census of Construction* (1997) for the particular contractor types.

5 The t-statistic for general contractors' capital expenditure per worker is 1.73 and significant only at the 10 per cent level. All other cases of statistical significance are at the 1 per cent level with adjusted R-squares ranging from 11 to 63 per cent.

6 Between 1967 and 1997, general contractors cut in half their share of construction worker employment (from 35 per cent to 24 per cent) while maintaining their share of material purchases at 42 per cent. Increasingly, general contractors are becoming jobbers rather than builders. Author's calculations are from *Census of Construction* – 1967 data are from Table B3, "Selected Statistics by Employment Size of Establishment with Payroll," while 1997 data are from "Detailed Statistics for Establishments With Payroll by Industry Group," Industry Series Summary.

7 These results are consistent with other analyses of productivity differentials between union and non-union construction, such as Allen's (1984).

8 Articles on skills shortages or labor shortages appeared regularly in *Engineering News Record* in the last half of the 1990s. A typical report is from April 1999: "Project safety tops construction managers' list of concerns, according to a survey of 65 CMs [construction managers]. . . . Close behind safety, at 35%, were finding and keeping quality workers at 33%, and litigation at 30%. Longer term, 35% see labor shortages as the main risk-management issue in the next 2 years, followed by safety, at 17%." "Safety and Labor Supply are Greatest CM Worries," *Engineering News Record*, 5 April 1999, 242 (13): 7.

9 Key papers regarding the construction cost effects of prevailing wage regulations are Fraundorf, Farrell and Mason 1984; Bilginsoy and Philips 2000; Keller and Hartman 2001; and Azari-Rad, Philips and Prus 2002.

REFERENCES

Allen, S. G. (1984) "Unionized construction workers are more productive," *Quarterly Journal of Economics* 99 (2): 251–74.

"Austin firm addresses labor shortages," *Texas Construction* (1998) 6 (12): 46.

Azari-Rad, H., Philips, P., and Prus, M. (2002) "Making hay when it rains: The effect of scale economies, seasonal and cyclical business patterns, and prevailing wage regulations on school construction costs," *Journal of Education Finance*, (Spring 2002): 997–1012.

Bilginsoy, C. and Philips, P. (2000) "Prevailing wage regulations and school construction costs: evidence from British Columbia," *Journal of Education Finance* (Winter 2000): 415–32.

Bureau of the Census (1967) *Economic Census, Construction, Industry Statistics*, "Selected Statistics by Employment Size of Establishments with Payroll," Table B2: 1B–4.

Bureau of the Census (1972–97, every 5 years) *Economic Census, Construction, Geographic Series*, "Detailed Statistics for Establishments with Payroll."

Bureau of the Census (1997) *Economic Census, Construction, Industry Series*, "Detailed Statistics for Establishments with Payroll by Industry Group," Table 4: 10–11.

Bureau of the Census, Microdata Access Branch (2001) *Current Population Surveys, Outgoing Rotations, 1979–2000,* produced and distributed on compact disk with extraction software by Unicon Research Corporation, Santa Monica, Calif.

Business Roundtable, Construction Cost Effectiveness Task Force (1997) "Confronting the skilled construction workforce shortage: A blueprint for the future," October, Washington: Business Roundtable.

Department of Labor, Bureau of Apprenticeship Training (1995) Electronic data provided to author.

Department of Labor, Bureau of Labor Statistics (2001) *Current Employment Statistics, State and Area.* Online. Available HTTP: http://data.bls.gov/cgi-bin/surveymost?sa (23 January 2002).

Fraundorf, M., Farrell, J. P., and Mason, R. (1984) "The effect of the Davis-Bacon Act on construction costs in rural areas," *The Review of Economics and Statistics* (February): 142–6.

Keller, E. C. and Hartman, W. T. (2001) "Prevailing wage rates: The effects on school construction costs, levels of taxation, and State reimbursements," *Journal of Education Finance* 27 (2): 713–28.

Koretz, G. (2000) "Why is hard hat pay falling?" *Business Week* 3683: 42.

Krizan, W. G. (1995) "Craft shortages creeping in," *Engineering News Record* 235 (26): 34.

Krizan, W. G. (1998) "Labor: industry trying to measure up," *Engineering News Record* 240 (26): 34.

Krizan, W. G. (2000) "Labor: costs may be hidden in hiring," *Engineering News Record* 244 (25): 96.

Londrigan, W. J. and Wise, J. B., III (1997) *Apprentice Training in Kentucky, A Comparison of Union and Nonunion Programs in the Building Trades,* Louisville, Ky.: Building Trades Apprenticeship Coordinators/Directors Association of Kentucky, Inc.

9 The United Kingdom

Privatization, fragmentation, and inflexible flexibilization in the UK construction industry

Mark Harvey

INTRODUCTION

While this book presents construction industries along a continuum of regulated and deregulated markets with a focus on the effects of deregulation, the construction industry in the United Kingdom presents a reality not of deregulation but rather of shifting regulation. State policy has shifted from the purchase of public housing units – a policy that stabilized construction demand – to a policy of encouraging home ownership through tax incentives – a policy that has exacerbated swings in demand by making demand more sensitive to interest rates. State policy has shifted from the direct employment of approximately 15 per cent of construction and related workers to the privatization of all public housing construction, maintenance, repair, and related workers. Changes in state taxation and social insurance policies triggered competitive pressures to shift employment for wages into self-employment as independent contractors.

The consequence of these shifts in state policy has been the explosion of mass self-employment in the construction industry. Collective bargaining has ceased to regulate the bulk of private sector construction. National employer organizations have been weakened. The proliferation of highly articulated and relatively chaotic subcontracting structures has reduced managerial control on the work site, and competitive pressures instituted by the state have forced almost all contractors – some willingly, others not – to shift to low-skill, low-wage, low-quality construction strategies. The UK presents a case where a shift in regulations rather than the notion of deregulation has directed the construction industry down the "low road" development path.

HISTORICAL CONTEXT

The construction industry in the United Kingdom has undergone major changes over the post-World War II period to what was already a quite distinctively organized sector of UK capitalism before the war. Parallel, if not necessarily precisely synchronous, changes took place in the product market, in the capital structure, and with employment relations.

In the housing construction market, three broad phases can be discerned. From 1945 to 1957/8, social housing dominated over private sector housing in terms of numbers of house completions. Between 1958 and 1977 there was a long period of rough parity between the public and private sectors in terms of housing output. Already presaged by retrenchment of public expenditure under the Labour government, the election of the Thatcher government in 1979 heralded a third phase, which included the retreat of the State, a sharp switch to home ownership, and much more volatile and cyclical swings in housing output and demand.

Very broadly parallel phases can be discerned in both the structure of the construction industry and the nature of construction employment. After a period of post-war renewal, a state-led period of mixed growth saw a rationalization of the industry, along with the growth of major and dominant national construction companies, combined with a significant if secondary public sector construction component. This was then followed by the increasing dominance of speculative construction companies, operating on a "design and build" basis, and of small subcontractors.

In employment terms, the first phase continued the patterns of employment and craft-based unions established in the pre-war period. In the central second phase, direct employment in both public and private sectors, with collectively bargained wages, craft and skill recognition, and training through apprenticeships, dominated. However, already during the latter stages of this phase, forms of "self-employment" began to emerge, especially following major industrial disputes in the commercial building sector in the City of London in the early 1970s. The third phase saw not only the massive privatization and reduction of the public sector employment, but also, even more significantly, the emergence of a distinctive form of self-employment as the pre-eminent form of "employment" in the private sector, the erosion of collective bargaining, and the undermining of skills and training. We shall now examine each of these three parallel aspects of change in order to account for the dynamics underlying changes in construction labor markets, and then review their consequences.

THE ECONOMIC EVOLUTION OF THE CONSTRUCTION MARKET

The evolution of the construction market from 1977–97 serves as a good descriptor for the evolution of the British economy as a whole. As can be seen from the changing shares of overall construction output in Figure 9.1, public housing descended from rough parity with private housing in 1977 to under 5 per cent in 1997, whereas private housing, showing considerable volatility, remained between 20 per cent and 25 per cent. Overall, the share of housing, private and

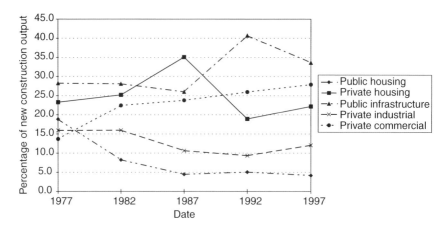

Figure 9.1 New construction, 1977–97, market evolution.

Source: Adapted from *Housing and Construction Statistics*, Annual Reports, 1987, 1997, Table 1.6.

public, in overall new construction fell from 42 per cent in 1977 to 26 per cent in 1997, a significant decline.

New construction for private industrial purposes downwardly followed manufacturing decline in the United Kingdom, while, conversely, the major gain in share of total construction output has been in private commercial development – i.e., offices and shops – reflecting the growth of the tertiary sector of the economy. Public infrastructure, especially road-building for private transport, also showed considerable gains. In Ball's review of construction output trends from 1955 to 1985, he noted the dramatic rise in public sector markets, both housing and other construction, from 1955 to 1972, as they doubled in value over this period, with the non-housing element trebling (Ball 1988). This latter showed faster growth than any other construction market. Thereafter, following the oil crisis, there is a sharp division between falling public sector demand across all spheres, and cyclical but growing private sector construction. The reshaping of the public–private divide was therefore a central feature of UK construction market demand over this period.

In its heyday, the State constituted a major client, a source of aggregated and concentrated demand, both at central and local levels. The shaping influence of this client/demand role of the State can be seen most clearly in the housing sector. Looked at from the standpoint of overall housing production, the first two phases can be seen as renewal and then expansion of the national housing stock. Only during this second phase, from 1952 to 1979, was there a sustained and continuous level of housing completions of over 250,000, taking growth in public and private sectors in total. Thereafter, the level of house completions declined until the 1990s, when it oscillated between historic low figures from

140,000 to mid-150,000. There has been an accelerating rate of aging of the housing stock. This was the Thatcherite "achievement."

The role of the State in housing construction was unique to the United Kingdom within the capitalist world. It helps to account for the specific structure of the industry in this country. The State, and in particular the local state, operated as construction client, employer, and landlord rolled into one. By the peak of public ownership of housing in 1986, the state directly owned 30 per cent of the total national housing stock (Emms 1990). The non-state social rented sector, by comparison, played a minor role until the progressive privatization of council housing, either by home ownership[1] or by transfer to private housing associations. Table 9.1 demonstrates the significance of this distinctively UK phenomenon.

For the construction industry, this central role of the State as client for social house building was crucial: demand was concentrated and had the same effect of "rationalization" of production that nationalization had in other sectors of the economy (Best 1990). There was a degree of concentration of demand not found in the private sector. This was most pronounced during the early 1960s, in the phase of the now decried "tower block mania" (Dunleavy 1981).

In this period of major housing renovation, 58 per cent of the total value of all housing contracts was accounted for by contracts larger than £500,000, of which 54 per cent were found in the public sector and only 4 per cent in the private sector (Dunleavy 1981: 37). The effect on the structure of the industry was equally significant. Seven firms accounted for 75 per cent of all industrialized high-rise flats in England and Wales, all state sector. Dunleavy has commented on the almost symbiotic relationship between local state clients and a particular major contractor. Effectively, state clientelism consolidated the role of major national housing construction companies.

During this second phase, when, for new housing construction, private sector contractors relied on the public sector for 40–50 per cent of their market, the local state also employed at a peak 200,000 building workers in Direct Labour Organisations, which are workforces employed by local government to undertake a range of work including construction, repairs and maintenance,

Table 9.1 Public sector and social rented housing, number of households, England and Wales

Year	Local (State) Authorities	Housing Associations	Public Sector as percentage of total
1914	20,000	50,000	29
1939	1,180,000	70,000	94
1953	2,400,000	90,000	96
1961	3,528,000	150,000	96
1971	4,803,000	200,000	96
1981	5,361,000	435,000	92
1988	4,569,000	543,000	89

Source: Emms 1990.

Note
Housing Associations are non-profit-making, non-state, "social rent" landowners.

highways, cleaning, and waste removal (Direct Labour Collective 1978, 1980). This amounted to approximately 15 per cent of the total construction labor force. Before the collapse of state social sector housing, these organizations played a significant and disproportionate role in relation to this percentage, setting standards of training. They were, however, the favorite target of Conservative governments both local and central and became the first victims of the Thatcherite program of privatization, piloting all subsequent legislation in this field.

The third phase witnessed a radical change in all aspects of the construction market, following first the oil crisis, which challenged the politics of the Labour Government, and then the Thatcherite program of privatization. Figure 9.2 demonstrates the main features of the change in the most sensitive area of housing construction.

From a period of virtual parity, the State ceased to be a major client for private contractors, while the role of Direct Labour Organisations was effectively terminated, at least in new construction, over the following 15 years. At the same time, there was a major restructuring of tenure, both through the progressive incentives for the sale of council housing to private ownership, and also, more significantly, by a major stimulation of home ownership through tax incentives and fiscal deregulation. In 1948, the United Kingdom had one of the lowest levels of home ownership among Western countries. However, successive Conservative governments had provided subsidies to home ownership that matched if not exceeded those to state social housing. By 1965, the UK had the second highest rate of home ownership of all Western countries, and by 1985 it was approaching parity with the US: 63 per cent by the mid-1980s and 67 per cent by the early 1990s (Forrest and Murie 1988; Ball, Harloe and Martens 1988). This process of

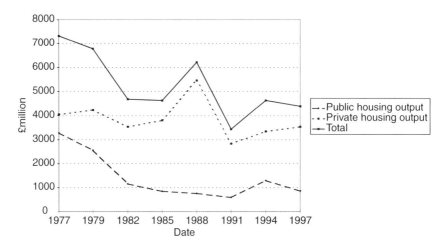

Figure 9.2 Public and private housing output, 1979–97, at 1985 constant prices.
Source: *Housing and Construction Statistics*, Annual Reports.

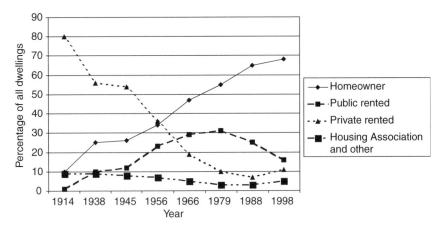

Figure 9.3 Housing tenure in Great Britain, 1914–98, percentage of all dwellings.
Source: *Housing and Construction Statistics*, 1980–2000.

shifting housing provision to almost exclusively private provision should not, therefore, be seen as simply "privatization" or "free market" operation. Home ownership is a testimony of massive fiscal regulation by the State, which changed the nature of housing demand (see Figure 9.3).

Consequently, the subsequent decades of this third phase have rendered the housing construction market particularly vulnerable to cyclical "highs" and "lows," while showing overall decline in new housing construction. Greater use of interest rate changes as a means of regulating money supply in Thatcherite economics acted as a negative or positive multiplier of demand, as more incomes became more vulnerable to interest rate change on mortgages. The housing construction market thus was subject to particularly violent swings, precisely because of this shift in housing tenure. In circumstances of high cyclical volatility in demand and house prices, the market induced profit-taking from asset speculation in land more than profit-making through the construction production process (Ball 1988). The Thatcherite decades became *par excellence* the decades of the speculative builder.

THE CHANGING STRUCTURE OF THE INDUSTRY

This particular market evolution produced manifest effects in the changing capital structure of UK construction industry, which is characterized by a concentration at the top by a few large companies and a plethora of small companies. In 1971, at a time when there was combined state- and private-led growth, 83 per cent of firms employed fewer than 13 direct employees and accounted for just 16 per cent of the total value of the output, whereas 0.1 per cent of firms with more than 1200 direct employees accounted for 24 per cent of the total value of output. In the following two decades, there was a quite dramatic restructuring of the construction industry, which can be seen in terms of share of employment and value of output by the largest and smallest categories of firms.[2]

There was a remarkable contrast between the rapid decline in direct employment by large firms and the equally rapid growth of directly employed workers by firms of fewer than 13 employees. In terms of overall numbers, the contrasting figures are striking: in 1971, firms of 13 and under employees employed 120,100 workers, against 174,100 by firms of over 1200 employees; by 1997, firms of 13 and under employed 136,100 (having peaked in 1983 at 168,300), while the largest firms now employed only 44,500. Figure 9.4 demonstrates the changing shares of total employment of direct employees, according to the size of the firm.

Clearly, there was a more than doubling of the share of direct employment by the smallest firms, from under 15 per cent to over 30 per cent, while, at the same time, there was a more than halving of the share of direct employment by the largest firms.

As remarkable, though, and perhaps more significant insofar as the effect cannot be attributed to the shift to self-employment, this change is closely mirrored by the share of total value output of firms during this period. Again, the share of total value of output of small firms doubled from 15 per cent to over 30 per cent, while the share of the largest firms fell by half, from just under 25 per cent to 12 per cent (see Figure 9.5).

The overall picture given by these statistics is one of rapid structural change, which could be described in terms of "flexibilization," including the downsizing of firms and outsourcing of functions. The industry had already been typified by very high levels of subcontracting. In 1982, before statistics ceased to register

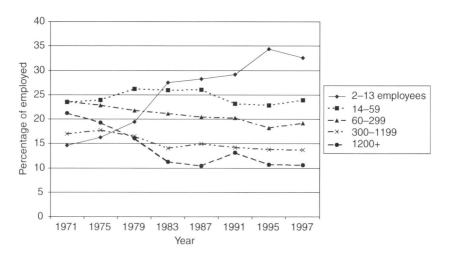

Figure 9.4 Size of firm by percentage of direct employees employed, 1971–97.
Source: *Housing and Construction Statistics*, Annual Reports, 1980–2000.

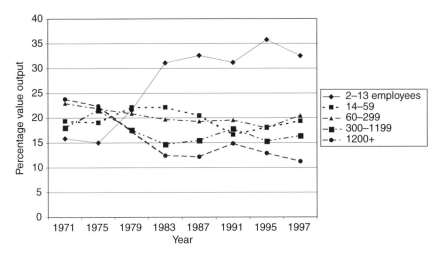

Figure 9.5 Size of firm by percentage of value output, 1971–97.
Source: *Housing and Construction Statistics*, 1980–2000, adapted from Table 3.3.

this aspect of the industry, firms employing more than 115 employees were subcontracting a third of the total value of their contracts, a figure which rose to 40 per cent for the largest firms (Ball 1988).

This quantitative picture is supported by a survey of private construction sites conducted by the author and others in the mid-1990s, called *Disparities in Wage Relations and Reproduction of Skills in the European Construction Industry* (Harvey 1995, 2000).[3] A survey of ten private sector sites and one public sector construction site in England in 1995 explored, as one aspect, the different patterns of contractual organization linked to different patterns of employment. There was a wide variety of sites, including, for example, airports, a church restoration, a new large scientific research establishment, and a police station. The range of main contract values was also broad, from £2.2 million to £600 million.

Three broad patterns were discerned, which illustrate how important it is to distinguish between different types of subcontracting, both in creating vertical chains and in the horizontal relations between subcontractors:

1 A flat hierarchy of subcontracting, with short vertical chains, often one tier only. Typically this involved a main contract management firm employing labor through a labor agency. It is in effect outsourcing labor, and as such can be seen as the self-employment alternative to direct employment. Given the statistics (see Figure 9.4), which show how major contractors especially divested themselves of direct employees, these can be seen as examples of the "hollowed-out firm."

2   Extended vertical chains of complementary capabilities at each tier in a vertical chain. This can be seen as "flexibilization without fragmentation." The dominant mode of contract management and a subcontracting system of organization of production are in operation, but with much repeat contracting and stable networks of subcontractors.

3   Extended vertical chains of subcontracting that parcel out contracts even of similar production content (for example, subdividing concrete work or dry-lining or electrical work into smaller packages). This can be characterized as "flexibilization with fragmentation." Extensive vertical chains are formed as successive parcels of work are subcontracted out, creating competitive and non-cooperative relations between subcontractors with similar capabilities at each tier of the vertical chain on a given production site.

Clearly, because of the small sample size of the survey, no conclusions can be drawn as to which of these three patterns is dominant in the United Kingdom, even in the knowledge that self-employment became the dominant employment form. It is significant, however, that major national contractors were represented in the sample, and they operated on a similar basis across all their contracts.

To illustrate and contrast these patterns of subcontracting, a relatively small contract of £6 million was divided into over 45 different subcontract parcels, many of which were further sublet by second-tier subcontractors on their own account. A high-level of turnover of contractors, and lack of any stable or regular contractual relations either between the main contractor and its subcontractors or between lower-level subcontractors in the chain, produced a highly unstable and uncoordinated project management. As many of the subcontract packages involved similar work content – e.g., carpentry and electrical work – competitive relations between on-site subcontractors with similar capabilities inhibited management coordination between them. Other sites involved as many as five tiers of subcontracting in the chain, with the main project management contractor having little control, or even knowledge, of the work of the subcontract firms below the second tier. Given the high levels of subcontracting in the United Kingdom, this kind of flexibilization with fragmentation was accepted as quite typical, even as a characteristic "natural" to construction production. Relative chaos became the norm.

Contrasted with this, an engineering construction site for a combined cycle gas turbine power station also involved several tiers in the subcontracting chain, but subcontracting of packages was related to specialization of capabilities of firms, and their complementarity in the overall production process. These subcontractors did not directly compete with each other for like work. Moreover, most were either subsidiaries of the main contractor or had repeated contracts over several years with the main contractor, and thereby had formed relatively stable networks. This can be seen as an example of flexibilization with coordination. Significantly, engineering construction here and elsewhere remains one of the few remaining bastions of direct employment, collective agreements, and shared project procedures.

By contrast, major civil engineering contractors[4] were typical of the "hollowed-out firm," directly employing next to no manual labor, but relying extensively on labor agency supply of self-employed labor. Road construction sites were the most representative of this type of operation, with an absence of extended tiers of subcontracting chains for a bulk of the contract. In some cases, the main contractor even owned the labor agency that exclusively supplied it with labor, and, in turn, the labor agency supplied labor exclusively to that contractor. In other cases, a "hollowed-out" main contractor sourced labor from a number of different labor agencies, who were offering different rates in competition with each other, and this led to much less stable labor supply relations between contractors and labor agencies.

Although no adequate surveys exist to establish the relative significance of these different types of contractual relations, it is fairly safe to assume that the two dominant forms to emerge in the United Kingdom from the mid-1970s are flexibilization with fragmentation and the hollowed-out firm reliant on agency labor.

LABOR MARKETS AND THE CENTRAL ROLE OF SELF-EMPLOYMENT IN UK CONSTRUCTION

The rise to dominance of self-employment from 1979 through 1995/6 is clear-cut. We shall now examine how regulatory changes have affected employment relations in marked ways.

There is arguably a strong linkage between type of firm organization and type of employment relation. Indeed, in one respect they amount to the same thing. From a formal contractual point of view, a self-employed operative is in fact an independent subcontracting enterprise. Thus, a construction site with 2000 self-employed workers technically and legally is a site with 2000 separate subcontracting firms, each with a distinct value parcel of the main contract. But at a more fundamental and realistic level, the weakening of the employment relation and contractual chain fragmentation, lack of inter-firm cooperation, and hollowing-out of firms can be seen as the result of a common process of how the construction market and competition within it have become instituted.

The construction workforce in the United Kingdom, including both employed and self-employed, although declining somewhat since the 1970s, still stands at 6–7 per cent of the total workforce, a proportion fairly typical of European countries. The UK, however, stands out since the election of the Conservative Government in 1979 as the country with the highest rate of growth of self-employment of any European country across all sectors, rising from 7.5 per cent in 1979 to 11 per cent in 1993, and now amounting to over 3 million self-employed in a workforce of 26 million. Moreover, even within this distinctive UK context, the construction industry was unique. In global terms, in 1993 the construction industry had 45 per cent self-employed of the total workforce, compared with the next highest industrial sectors: distribution, hotels, and catering at 14 per cent; transport and communications at 13 per cent; and banking, insurance, and finance also at 13 per cent (*Labour Market Trends* 1999).

Yet this gives only a partial picture, because estimates of self-employment are notoriously conservative as a result of the known difficulties of registering transient, and, as will be seen, essentially illegal workers.[5] Within the private sector, it is notable that self-employment is especially a phenomenon of manual workers, with technical and supervisory staff still overwhelmingly directly employed. It should also be borne in mind that the construction industry workforce incorporates many small independent artisans undertaking repairs and maintenance and small-scale construction – in value terms a significant share of total output. These can be assumed to figure in the workforce figures as a reasonably constant percentage throughout the period. In contrast, in continental European countries where independent artisans are legally more restricted to the category of small-scale jobbing construction, their typical percentage of the construction workforce amounts to little more than 10 per cent. The truly significant change reflected in Figure 9.6 therefore represents mainly the growth of manual self-employment on private construction sites.

The broad picture, therefore, is one of substantial increase of self-employment, from under 30 per cent in 1977 to a peak in 1995 of over 60 per cent before a decline again to just over 50 per cent in 1999. There is no doubt that a massive change took place in the construction labor market during this period. At this point, it is worth emphasizing some of the main implications for those who shifted out of direct employment into self-employment. Because of the change in status, these changes occurred:

- Payments for these workers automatically fell outside the scope of any possible wage bargaining, as technically they were no longer paid wages.

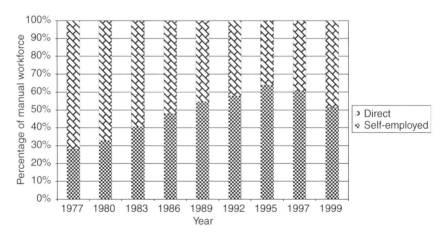

Figure 9.6 Private sector manual workforce, employed and self-employed, 1977–99.
Source: *Housing and Construction Statistics*, 1987–97, Table 2.1.

- They lost entitlement, during most of this period, to holiday pay, sick pay, unemployment benefit, and, in most cases, any form of pension arrangements.
- Automatically too, they lost any of the normal employment protections for dismissal or discipline that form part of a collective agreement related to direct employment.

Although before the 1980s large numbers of construction workers were similarly vulnerable, this period became one of mass self-employment. No other group of workers, even in the "deregulated" labor markets introduced by Thatcher, experienced a similar degree of vulnerability on such a scale.

Far from being a deregulated market, however, for reasons that we shall now explore, the market price for different employment statuses was set by the State through the taxation and national insurance systems in a way unique to the construction industry. This triggered a state-induced competitive pressure to shift labor from employment into self-employment, producing a downward spiral not only in the standards of employment enjoyed by construction workers, but more generally in the productive capacity to organize and manage labor. The reproduction of improvement skills, which might enable or stimulate the adoption of new technologies, was also impaired.

In order to comprehend the specifics of the UK labor market, it is worth underlining the fact that the distinction between employment and self-employment rests on customary case law, rather than statutory legal status. The distinction is based on criteria for judging whether there is a "contract of service" or a "contract for services." In the absence of a statutory framework, one of the central regulatory frameworks shaping the UK labor market has been the taxation system. Taxation regimes in the United Kingdom assume a particular importance, and quite distinctively so for the construction industry, where a unique system of taxation associated with employment status has evolved (Harvey 1995). Allied to the legal distinction between "contract of service" and "contract for services," the taxation and national insurance systems effectively created a two-price labor market.

The result is a major institutional discrepancy or contradiction between status as ascribed by the taxation system and that designated by case law criteria. It was generally accepted by officials in the Inland Revenue and National Insurance departments (interviewed in 1995) that at least 80 per cent of those designated as self-employed under the tax system would not meet the criteria laid down by case law. One can argue that there is now mass illegal self-employment, tolerated by the State, inasmuch as most self-employed do not determine their own working hours, supply their own equipment or materials, allocate or designate their own work, set their own rate of pay, or submit invoices for work undertaken. In short, they manifest none of the legal criteria necessary to establish self-employment (Burchell, Deakin and Honey 1999).

In the construction industry, access to the tax status of self-employed has, until very recently, been facilitated by two unique tax employment categories: the 714, where "employees" are responsible for submitting their own tax returns, and the SC60, or subcontract 60, where "employees" are taxed at source at the standard rate of income tax. The SC60 form of self-employment became by far the largest category and is perhaps the most extraordinary, inasmuch as for tax purposes these workers effectively paid the same rates of tax as the employed, and, in the manner of the pay as you earn (PAYE) income tax, this was deducted by the contractor prior to subcontracting. For this reason, tax officials have described it as "quasi-PAYE for quasi-employment." Access to the 714 was supposedly much more tightly regulated, entailing the issue of an identity card, on condition of proof of business address, and previous tax record. In practice, as interviews with staff of the Inland Revenue Department and the Department of Social Security confirmed, the 714 certificate, although more difficult to obtain than the SC60, was easily available without proof of a genuine, independent, self-employed economic activity under a "contract for services."

The categorization of employment status by taxation regime rather than legal criteria, however, has a major and intrinsic impact on the market price of labor (Harvey 2000). The effect of self-employment designation is to remove the employer's obligation to pay any national insurance contribution, amounting at the peak of self-employment to an immediate and direct 12.2 per cent saving on labor costs. Additionally, workers have reduced payments for national insurance (in line with reduced claims on all but means-tested state benefits), so lowering the supply-side cost of labor. These immediate and direct effects alone produced a two-tier labor market between direct employment and self-employment. Moreover, employers no longer paid holiday pay or any other leave entitlements. It has been therefore estimated that the overall reduction in labor costs through self-employment amounted to between 20 and 30 per cent (Evans and Lewis 1989).

A tax and insurance regime of this sort therefore instituted an irresistible competitive pressure on employers to shed direct employees. Given the dubious legal status involved in mass self-employed, it is well known that a culture of tax evasion of both direct and indirect tax became the established norm within the industry. It can therefore be argued that the overall cost to national insurance and tax revenue of mass self-employment on the scale that developed was, at its height, between £2 and £4 billion per annum (Harvey 1995). This amounts to a major state subsidy to a degraded form of employment within the construction industry.

In stressing the societal significance of taxation and insurance regimes as institutional characteristics of labor markets, and thus the uniqueness and specificity of pricing mechanisms and their regulatory effects in different countries, it is equally important to emphasize that taxation and insurance regimes do not fall out of the skies. Rather, they develop and evolve in interaction with those markets. It is important, when dealing with legal and fiscal frameworks, to treat them as part of dynamic change, rather than as the static context within which "economic" dynamics occur. Many factors are involved in these developments: political ideologies of labor markets, institutional rivalries and power struggles between different taxation and insurance departments, economic cycles enmeshed with speculative building and interest rate controls, the contradiction between legal and taxation institutions, and so on. The changes in taxation regimes now described thus closely parallel the phases analyzed above for changes in the construction product market and firm organization.

The present UK construction "taxation regime"[6] has also evolved over the past three decades. While a longer-term historical perspective is outside the scope of this paper (see Austrin 1979, 1980; Rainbird and Syben 1991; Birch 1993), the present taxation regime has stimulated the emergence of new forms of self-employment. It too was in part a response to the effects of previous tax regimes.

Broadly, four periods can be discerned:

- 1960–70: A period of major industrial strife, especially in the City of London, where unionization and union power reached a high point. This period stirred employers and government to contemplate the encouragement of self-employment as a means of counterbalancing union power.
- 1970–80: The early establishment of the self-employment taxation regime, the 714 and SC60 system. The 1970 Conservative government initially instituted a semi-regulated taxation regime for the construction industry that laid the basis of the taxation regime for the next 25 years. Rapid expansion of self-employment followed.
- 1980–95: The implementation of incentives for self-employment. The Thatcher government relaxed criteria for registering as self-employed by an administrative policy decision in 1980 (Officials, Technical Support Unit of the Contributions Agency, interviews by author, 1994–5).[7] With tax and insurance incentives in place, the self-employed workforce has more than doubled in proportion to the total workforce, from under 30 per cent of manual employees in 1980 to 62 per cent in 1995.
- 1997–present: The partial reimposition of barriers to false self-employment. In 1997, the Inland Revenue and Contributions Agency sought to reimpose on the construction industry the same employment criteria that applied to any other industry. This has met with limited success with 185,000–210,000 workers shifting into employment. The most recent signs, however, are that, following the introduction of the Construction Industry Scheme (CIS) in August 1999, that shift has been reversed (Harvey 2001).

These institutional vicissitudes in taxation and insurance regimes over the past two decades are reflected quite sharply in Figure 9.7, which shows rapid shifts of labor from direct to self-employed and back in response to regulatory changes. The central decade is one where a continuous shift of around 10 per cent per annum of the workforce from direct to self-employment was interrupted only by periods of recession in the construction industry, where it is likely that self-employed were more rapidly shed than the directly employed (i.e., in 1983, and in 1989/90). At the beginning and end, the effects of the introduction of "the policy" and its reversal are clearly manifest. Finally, there is some suggestion that

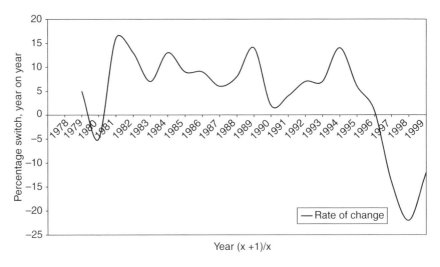

Figure 9.7 Rate of switch from direct to self-employment, year on year, 1979–99.
Source: Adapted from *Housing and Construction Statistics*, 1987–97, Table 2.1.

the introduction of the CIS has halted the dramatic imposition of a tighter definition of self-employment in 1995.

THE CONSEQUENCES OF MASS SELF-EMPLOYMENT

In this section, we shall briefly review the consequences of mass self-employment on trades union membership, collective bargaining, and wage fluctuation and dispersion; on training and apprenticeship; on health and safety; and on self-employed workers' own experience of self-employment and growing insecurity.

Trade union membership, collective bargaining, and wage fluctuation

The first consequence of the shift to mass self-employment was that hundreds of thousands of construction workers ceased to be within the orbit of collective wage bargaining. The major construction trades unions – Union of Construction Allied Trades and Technicians (UCATT), Transport and General Workers Union (TGWU), the GMB, Electrical Engineering and Plumbers Union (EPTU), and Amalgamated Engineering Union (AEU) – refused to consider recruiting members from amongst the self-employed through most of the 1980s as they were considered, *ipso facto*, to undermine union principles. As a consequence, by 1999 there was a marked contrast between public sector union density, where direct employment still obtained, and the private sector, where it remained only in small enclaves within the sea of self-employment. The former had a union density of 75 per cent; the latter, only 14 per cent (Cully *et al.* 1999; *Labor Market Trends* 1999; Millward, Bryson and Forth 2000). Overall, in public and private sector, 55 per cent of workplaces had no union present and fell outside collective agreements,[8] and in a further 8 per cent of workplaces unions were present but without recognition (Cully *et al.* 1999). Clearly, these figures would be more pronounced for the private sector alone.

By a number of criteria, the results of the Workplace Employee Relations Survey (WERS) and the Workplace Industrial Relations Survey (WIRS) show the construction industry to occupy an extreme position in UK industrial relations (Cully *et al.* 1999; Millward, Bryson and Forth 2000). Of all standard industrial sectors, construction has the lowest level of employee participation in representative bodies; the highest number of workplaces without a representatives committee, 73 per cent; and the highest percentage of workers working more than 48 hours per week, 32 per cent (Cully *et al.* 1999, Table 4.6). After two decades of declining union membership, at 30 per cent union density in 1998, construction was lower than all other main primary production sectors. Compare this union density, for example, to that of energy and water supply at 68 per cent; extraction industries and chemicals at 47 per cent; metal and engineering at 45 per cent; other manufacturing at 33 per cent; and transport and communications at 54 per cent (Millward, Bryson and Forth 2000).

The *Disparities* survey cited previously in this chapter explored the effect of self-employment on wages of the self-employed, and, although it was small scale, with a sample amounting to 3043 self-employed on 12 construction sites, it provided otherwise unavailable information. First, it should be noted that the form of payment is not the hourly rated pay common to directly employed workers, but payment by the shift, normally a 10- to 12-hour shift, and a 6-day week (Harvey 1999).[9] Two further notable characteristics distinguished self-employed wages: the degree of wage dispersion and the degree of wage fluctuation during the economic cycle. In the major craft union collective agreements covering manual operatives, wage differentials from bottom to top have a range of 100:126 under the Building and Civil Engineering agreement or 100:167 under the Engineering Construction agreement, with the lowest paid laborer indexed at 100. Taking all sites of the *Disparities* survey, the range of wage differential between lowest and highest paid self-employed worker is staggeringly greater, at a 100:299 ratio. There are also great variations within sites, ranging from 100:179 to 100:211. Even within a single narrow category of worker, such as "shuttering carpenter," price per shift varied by 15–40 per cent on the same site. Thus, it appears in the most "deregulated" markets, the presence of fragmented subcontracting and self-employment impedes the emergence of a homogeneous equilibrium price for labor. Moreover, absence of any normalized relation between price and different levels of skill significantly reduces the incentives for training.

In terms of wage fluctuations, workers were asked whether and to what extent their gross nominal wage had varied over the 5 years preceding 1995, which included the recession of 1989–91. The contrast between the directly employed minority (mostly within the public sector) and the self-employed majority was marked. Whereas 66 per cent of the directly employed enjoyed relative wage stability with nominal wages keeping in line or just below inflation, 74–75 per cent of self-employed suffered wage cuts, and between a fifth and a quarter experienced cuts of gross nominal wage above 30 per cent, sometimes as much as 50 per cent.

Overall, therefore, whereas the formalities of national collective wage bargaining continue to be observed and to have limited application, trades unions have effectively lost negotiating powers in the bulk of the private sector. Their main activities now focus on the defense of employment rights, industrial tribunals, and health and safety campaigns.

While it might be expected that employers gained relative power as a result of the major changes in labor market institutions and weakening of the trades unions, it is by no means clear that there is a simple zero-sum game, either in wage determination or in managerial control. The major national employer organizations,[10] which paralleled the trades union bargaining structures, were also in important respects weakened by the changes. In terms of managerial control, the dominant patterns of subcontracting described above in many ways manifest a loss of authority, and, in terms of wage determination, large firms were powerless to resist the competitive pressures instituted by the State, forcing them down a low-cost, low-quality employment road willingly or unwillingly.

The effect on training and the absence of skill recognition

The hollowing-out of the firm and the shift by major contractors from the engagement of direct employees to self-employed had necessary implications for skill reproduction. Apprenticeship as a system of skill reproduction implies employer responsibility for the acquisition of skills by young workers. With the breaking of the employer–employee relationship, that responsibility evaporated. Before the *Housing and Construction Statistics* stopped publishing figures on trainees in the workforce, the number of all trainees (of whatever level or training duration) in the industry had fallen from 62,800 in 1977 to 48,000 in 1986. The decline in the number of apprenticeships, which rely on stable employment relations, was even starker, from 8,700 registered Construction Industry Training Board apprentices in 1985 to 2500 in 1994 (CITB 1994). In their comparative study of training and quality in the construction process, Clarke and Wall (1996) noted that on-site apprentice training was nonexistent among employers engaging the self-employed, a feature also confirmed by the *Disparities* survey. While the National Vocational Qualification (NVQ) system was designed to modernize and upgrade the traditional apprenticeship system, and to remain essentially rooted in a combination of workplace-based training and formal college training, the effect of self-employment has been to undermine almost entirely workplace training across the vast swath of the private construction industry. Their study discovered that over 90 per cent of 1400 construction trainees in a large London Further Education College were either unemployed or self-employed, without the support of employer and on-site training. In those circumstances, it was unsurprising that they discovered the ratio of trainees to craftworkers was as much as half that of Germany and the Netherlands.

A less direct indication of the effect of self-employment is the absence of the development of skills and skills recognition in a key area of modern building technology: prefabrication and concreting. It is remarkable that workers engaged in concreting have no recognized skill status. In some cases, as with the "shuttering carpenter," this could be interpreted as a Taylorist process of de-skilling and skill fragmentation. However, for concrete pourers, finishers, or steel fixers, that interpretation cannot hold, as there is no subdivision of an existing craft. Amongst the self-employed, informal categories have emerged, such as nip-hands, and semi-skilled and skilled steel fixers. However, no formal recognition or training supports the development of these skills. There is a strong argument that the absence of employer responsibility for training, through the decades of mass self-employment, has resulted in an overall failure to develop a coherent bundle and organization of skills and levels of skills related to a key area in modern construction, in marked contrast to continental Europe. Other areas, such as floor-laying or dry-lining, reflect a similar emergence of a narrow task-related skill. There is a correspondence between the emergence of small labor-only contractors and labor agencies as the two dominant forms of self-employed labor supply and this restrictively narrow form of skill development.

Self-employment and the threat to life and health

There have been no systematic studies of the impact of deregulation and the growth of self-employment on health and safety within the construction industry. Such analysis is hampered by the fact that many of those fatally or seriously injured while being registered self-employed are deemed after the event to have been employed for purposes of compensation.

Nonetheless, the fatal and major accident rate in the construction industry remains at very high levels in the United Kingdom, with the number per 100,000 ranging from 223 in 1995/6 to 411 in 1996/7 (*Health and Safety Executive* 2001a). This accident rate is approximately four times higher than that for all industries. There have been 70–100 deaths in every year since 1991, with a total construction workforce including self-employed of approximately 1 million. There is some evidence that, with bogus self-employment no longer declining as rapidly, fatalities have once more begun to rise, with 92 in the year 1999/2000, their highest level for more than a decade (*Health and Safety Executive* 2001b). A recent government industry conference identified casual employment as being at least in part responsible for this unacceptably high-level of accidents (*Financial Times*, 27 March 2001; UCATT 2001).

Although health and safety is ultimately the responsibility of the main contractor on any given construction site, there is no doubt that the loss of managerial control resulting from multiple tiers of subcontracting seriously fragments that responsibility.

The experience of self-employment and growing insecurity

The impact on construction workers of the shift to self-employment in terms of employment rights has already been noted. It remains perhaps to further emphasize that this loss of rights leads to a vicious spiral in which, either through injury or old age, these workers face a future where their rights to benefits and pensions will be largely limited to means-tested benefits. This is particularly significant in view of the fact that this form of false self-employment carries with it much higher levels of employment insecurity.

The *Disparities* survey demonstrated that by far the most vulnerable were the workers under the dominant form of self-employment, the SC60 workers. Table 9.2 shows that that the average length of service of an SC60 worker with the current employer, at 1.19 years, is less than a third of that of a holder of a 714 certificate and approximately a sixth of that of a direct employee. There is a clear hierarchy of insecurity of employment, which is also reflected in the average lengths of employment over a working life. The table also shows that SC60 workers also experienced by far the highest rates of unemployment between jobs, with 56 per cent of SC60 workers suffering frequent periods without work (and without unemployment benefit).

This picture is confirmed if the same data are analyzed in terms of the employers of self-employed labor. Agencies that supply the large "hollowed-out" firms are the institutional conduit for the short-term, contingent employment, as one might expect within the UK regulatory framework. Thus, 69 per cent of labor supplied by agencies have an average length of service of 6 months or less, as against 24 per cent employed by Labour Only Subcontractors – that is, subcontractors that provide only labor to the construction project and no materials. Conversely, none of the agency-employed workers had more than a year with an

Table 9.2 Stability of employment for direct employees compared to two categories of self-employed workers, the 714 and the SC60

	Employment status		
	Direct	714	SC60
Average length of service with current employer	6.68 years	3.93 years	1.19 years
Average years with employer during lifetime	3.3 years	2.5 years	1.2 years
Periods of unemployment (as % of status)			
None	69	42	22
Few	18	16	22
Frequent	13	42	56

employer, as against 41 per cent employed by subcontractors. It is therefore important to recognize that even within self-employment, there are significant institutional levels and extremes of insecurity and contingency, articulated by the SC60 employment status on the one hand, and by labor agencies on the other.

CONCLUSION

A catch-phrase summary of the changes in the UK construction industry over the past three decades would be "inflexible flexibilization." Both in terms of firm organization and management of production, and in terms of employment relations, outsourcing has involved fragmentation and loss of coordination rather than more efficient systems of resource allocation and control of production. The result has therefore increased rather than decreased constraints on enhancing productivity or adopting new technology.

Whether through changing institutional characteristics of product demand, and the growth of speculative housing and commercial markets, or through the emergence of mass self-employment, the recent history of the UK construction industry illustrates distinctive processes of institutional economic transformation. Both product and employment markets have been fundamentally shaped by the changing role of the State, rather than by a reduction in the role of the State. The retreat of the State as major direct client in the product market was complemented by the state promotion of home ownership through fiscal incentives and financial regulatory frameworks. The decline of the (local) state as a significant employer of construction work and the weakening of employment protection was counterbalanced by the creation of a two-tier labor market, and a massive subsidy for self-employment through evolving taxation and insurance regimes. Thus, when exploring societal specifics, in the construction or any other industry, it is important to recognize that analysis of the dynamics of instituted economic processes should replace analysis in terms of a dichotomy between "economic" market forces on the one hand and "societal" regulatory frameworks on the other.

NOTES

1 Initiated first by the Conservative MacMillan government in 1952, restricted by the Labour government of 1967/8, and then given full support by the Thatcher Housing Act 1980, the Housing and Building Control Act 1984, and the Housing and Planning Act 1986 (Forrest and Murie 1988).

2 A word of caution at this point: The official statistics classify the size firms in terms of numbers of direct employees. Thus, over a period when there were progressively fewer direct employees overall, size of firm by number of employees does not reflect accurately the changing structure of industry. Thus, there could be fewer firms with large number of direct employees, or firms could be officially shrinking in size, but in reality these firms were controlling the economic activity of larger numbers of self-employed operatives.

3 This research project took place in 1995–6, funded by the Leverhulme Foundation and based in the Education, Training and Labour Market Research Group at the University of Westminster. I am grateful for the collaboration with the other researchers involved, Linda Clarke, Frédéric Gérardin, Rachid Belkacem and Jörn Janssen.
4 In the North American context, civil engineering contractors are often referred to as "heavy-and-highway contractors."
5 The figures are generally recognized to underestimate self-employment by at least 200,000 (Cannon 1994).
6 "Taxation regime" comprises all levy forms, taxation properly speaking, national insurance, and – although outside the scope of this paper – training levies, which are central to the reproduction of skills within the construction industry, but which are industry-based rather than state-based.
7 The policy is set out in Administration, Special Cases, CC 16286, where it states "there is no equivalent Department of Social Security legislation on subcontractors in the construction industry.... Since 1980, and as a matter of administrative policy, such workers are accepted as self-employed for National Insurance purposes, provided they have a 714 or SC60 certificate. This normalization of self-employed status based on self-declaration is at the root of the conflict of the social dumping of UK construction workers in Europe.
8 In the United Kingdom, if the union is not recognized by the firm, collective agreements do not apply.
9 The introduction of the Working Time Directive has had little impact on the self-employed as they are excluded from its remit by their employment status.
10 There are employer federations for the different construction sectors, and in 1997 there was further consolidation into the Construction Confederation, with 5000 companies representing 75 per cent of construction turnover.

REFERENCES

Austrin, T. (1979) *Industrial Relations in the Construction Industry: some Sociological Considerations on Wage Contracts and Trades Unionism, 1919–1973*, PhD Thesis, University of Bristol.
Austrin, T. (1980) "The 'Lump' in the UK Construction Industry," in T. Nichols (ed.), *Capital and Labour: Studies in the Capitalist Labour Process*, London: Fontana.
Ball, M. (1988) *Rebuilding Construction: Economic Change in the British Construction Industry*, London: Routledge.
Ball, M., Harloe, M., and Martens, M. (1988) *Housing and Social Change in Europe and the USA*, London: Routledge.
Best, M. H. (1990) *The New Competition: Institutions of Industrial Restructuring*, Cambridge: Harvard University Press.
Birch, J. (1993) *Employed or Self-Employed? Contract Labour in the British Construction Industry*, Working Paper, Geneva: International Labour Office.
Burchell, B., Deakin, S., and Honey, S. (1999) *The Employment Status of Individuals in Non-Standard Employment*, Report for the Department of Trade and Industry, March, London: HMSO.
Cannon, J. (1994) "Lies and construction statistics," *Construction Management and Economics* 12: 307–13.
Clarke, L. and Wall, C. (1996) *Skills and the Construction Process: A Comparative Study of Vocational Training and Quality in Social House Building*, The Joseph Rowntree Foundation, Bristol: Policy Press.

Construction Industry Training Board (CITB) (1994) *Annual Report*, Norfolk: CITB.

Cully, M., Woodland, S., O'Reilly, A. and Dix, G. (1999) *Britain at Work: as Depicted by the 1998 Workplace Employee Relations Survey*, London: Routledge.

Direct Labour Collective (1978) *Building with Direct Labour*, London: Conference of Socialist Economists.

Direct Labour Collective (1980) *Direct Labour Under Attack*, London: Conference of Socialist Economists.

Dunleavy, P. (1981) *The Politics of Mass Housing in Britain, 1945–1975*, Oxford: Clarendon Press.

Emms, P. (1990) *Social Housing: A European Dilemma*, Bristol: School for Advanced Urban Studies, University of Bristol.

Evans, S. and Lewis, R. (1989) "Destructuring and de-regulation in the construction industry," in S. Tailby and C. Whitstone (eds), *Manufacturing Change: Industrial Relations and Restructuring*, Oxford: Blackwell.

Forrest, R. and Murie, A. (1988) *Selling the Welfare State: The Privatisation of Public Housing*, London: Routledge.

Harvey, M. (1995) *Towards the Insecurity Society: The Tax Trap of Self-employment*, London: The Institute of Employment Rights.

Harvey, M. (1999) "Economies of Time: A framework for analysing the restructuring of employment relations," in A. Felstead and N. Jewson (eds), *Global Trends in Flexible Labour*, London: Macmillan, 21–41.

Harvey, M. (2000) "Systemic competition between high and low 'social cost' labour: a case study of the UK construction industry," in L. Clarke, P. de Gijsel, and J. Janssen (eds), *The Dynamics of Wage Relations in the New Europe*, Dordrecht: Kluwer Academic Publishers.

Harvey, M. (2001) *Undermining Construction: The Corrosive Effects of False Self-employment*, London: Institute of Employment Rights.

Health and Safety Executive (2001a) "Key facts: Injuries in the construction industry 1996/7 to 1999/00," http://www.hse.gov.uk.

Health and Safety Executive (2001b) "Annual Health and Safety Statistics, 2000/2001," http://www.hse.gov.uk/statistics/2001/hsspt1.pdf.

Housing and Construction Statistics, Annual Reports (1980–2000), London: HMSO.

Labour Market Trends (1999) Office for National Statistics, Vol. 107, London: The Stationery Office.

Millward, N., Bryson, A., and Forth, J. (2000) *All Change at Work? British Employment Relations 1980 to 1998*, London: Routledge.

Rainbird, H. and Syben, G. (1991) *Restructuring a Traditional Industry*, New York: Berg.

Union of Construction Allied Trades and Technicians (UCATT) (2001) *Construction Safety: Building a New Culture*, London: UCATT.

10 Republic of Korea

The drivers for change in the Korean construction industry: regulation and deregulation

Jin Ho Yoon and Byung-Goo Kang

INTRODUCTION

The Korean construction industry has maintained high growth along with rapid growth of the whole economy over the past 40 years. However, it received a severe blow from the economic crisis at the end of 1997, after which construction output decreased by 20 per cent. As a result, the number of employed in the construction industry decreased from two million in 1997 to less than one and a half million in 1999. Moreover, the productivity of the construction industry is deteriorating because qualified workers are not willing to enter the industry. Low-skilled labor prevails in the Korean construction market. Nowadays construction workers suffer from mass unemployment, job insecurity, poor working conditions, low wages, and long work hours.

It is important to understand that, although the bankruptcy of many construction firms after the economic crisis is the direct cause of this decline, the fundamental sources of decline are rooted in the structural problems of the industry.

The Korean case shows an interesting mix of regulation and deregulation in the construction industry and its labor market. For a long time, the Korean government has maintained strong regulations on the construction product market, including high entry and mobility barriers. These regulations brought about various problems such as an oligopolistic market structure, collusion and corruption among big companies, the problems inherent in a multi-layer subcontracting system, and poor quality of construction works. Contrastingly, there is almost no regulation on the construction labor market, either by the government or by the trade unions and collective bargaining. This lack of regulation results in numerous poor outcomes in the construction labor market, such as unstable employment, low wages, lack of training, and exclusion from social security systems.

Recent deregulation policies have worsened the situation considerably. New deregulation in the product market has led to fierce competition, but the mechanism for fair competition is not well developed yet. On the other hand, the unregulated construction labor market does little to protect construction workers. Unstable employment has become widespread, the role of illegal temporary agencies has expanded, and regulation on industrial safety and health has been weakened. Thus, the Korean case shows the worst combination of regulation and deregulation.

THE STRUCTURE OF THE KOREAN CONSTRUCTION INDUSTRY AND GOVERNMENT REGULATIONS

The development of the construction industry

The Korean construction industry has grown rapidly along with high economic growth since the 1960s. Figure 10.1 shows that, although the Korean construction industry has experienced a number of business cycles, the growth rate has been very high before 1997 except during a few years.

Table 10.1 shows that the construction output, measured in 1995 constant price, increased to 46 trillion won in 1997, which was more than ten times the output of 1970. However, the growth of the construction industry stagnated in the 1990s with the saturation of demand. The share of construction output increased from 7.4 per cent of GDP in 1970 to 10.9 per cent in 1997, but dropped to 8.7 per cent in 1999. Also, the growth of construction investment stagnated through the 1990s and dropped rapidly immediately after the economic crisis of 1997. After increasing from 12.6 per cent of GDP in 1970 to 22.8 per cent in 1990, the share of construction investment declined to 16.5 per cent of GDP in 1999. The share of employees in the construction industry increased from 2.9 per cent of total employees in 1970 to 9.5 per cent in 1997, but it dropped to 7.3 per cent in 1999. Together, these data indicate both a sharp

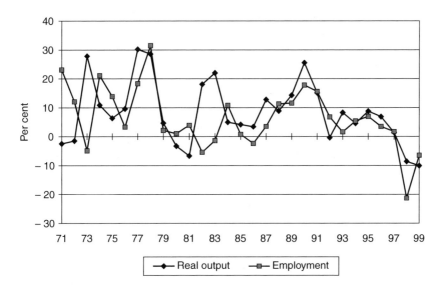

Figure 10.1 Growth rates of construction outputs and employment.

Table 10.1 The share of construction industry in the Korean economy

	Real figures* (Trillion won, thousand persons)					Annual growth rate (%)		
	1970	1980	1990	1997	1999	70–79	80–89	90–99
GDP	56	115	263	423	436	8.6	7.6	6.2
Production								
Construction	4	11	30	46	38	12.7	7.9	5.2
% of GDP	7.4	9.8	11.4	10.9	8.7	–	–	–
Investment								
Construction	7	19	60	89	72	9.2	13.4	5.2
% of GDP	12.6	16.7	22.8	21.1	16.5	–	–	–
Employees								
Total	9,618	13,683	18,085	21,106	20,281	3.8	2.6	1.5
Construction	281	843	1,346	2,004	1,476	13.3	3.3	3.2
% of Total	2.9	6.2	7.4	9.5	7.3	–	–	–

Note
* 1995 constant price.

Sources: Ministry of Construction and Transportation, *Yearbook of Construction & Transportation Statistics*, various years; Korea National Statistical Office, *Annual Report on the Economically Active Population Survey*, various years; Korea National Statistical Office, *Report on the Construction Work Survey*, various years.

absolute and a relative decline in Korean construction since the 1997 economic crisis.

The Korean economy was almost in full employment before the 1997 economic crisis. The unemployment rate was just 2.0 per cent in 1996. Construction was also at full employment. The number of construction workers increased rapidly as construction investment continuously expanded in the 1970s and 1980s. While the demand for construction workers increased in accordance with prosperous conditions of the construction market, new workers did not proportionately flow into the market, because of low wages and poor working conditions. Therefore, the Korean construction industry was confronted with labor shortages after the middle of the 1980s (Kim 1997). Now construction workers are in excess supply because of the crisis in the industry since 1997. Unemployment increased sharply due to the bankruptcy of many construction firms and redundancy dismissals since the economic crisis. The number of employed in the construction industry decreased from 2,004,000 in 1997 to 1,476,000 in 1999. Table 10.2 shows that the number of the unemployed in the construction industry increased from 50,000 in 1997 to 219,000 in 1999. This dramatic increase in unemployed construction workers accounted for 20.8 per cent of total exit unemployment in the overall Korean economy. The exit unemployment rate in the construction industry increased from 2.4 per cent in 1997 to 12.9 per cent in 1999, 2.5 times higher than the unemployment rate in mining and manufacturing.

Table 10.2 The exit unemployment rate by industry, in thousands of persons and per cent

Year	Number unemployed (Rate)	Exit unemployment		
		All industries	Mining & manufacturing	Construction
1985	622 (4.0)	414 (2.7)	127 (3.4)	108 (10.6)
1990	454 (2.4)	240 (1.3)	77 (1.5)	37 (2.7)
1995	419 (2.0)	255 (1.2)	69 (1.4)	38 (2.0)
1997	556 (2.6)	318 (1.5)	80 (1.7)	50 (2.4)
1998	1,461 (6.8)	1,190 (5.6)	279 (6.6)	298 (15.9)
1999	1,353 (6.3)	1,054 (4.9)	194 (4.6)	219 (12.9)

Source: Korea National Statistical Office, *Annual Report on the Economically Active Population Survey*, various years.

Note
Rate of exit unemployment in j industry = unemployment from j industry ÷ (employment in j industry + unemployment from j industry).

The regulations on the construction industry

Since the 1960s, the Korean government has maintained strong regulations on the construction industry. Particularly, the government has used entry and mobility barriers as the principal means of regulations on the construction industry, which results in an oligopolistic construction market in the country. This type of regulation is quite exceptional among the countries in the Organisation for Economic Co-operation and Development (OECD). There have been three kinds of regulations that have acted as entry and mobility barriers in the construction industry.

Business licenses

The first regulation was the construction business license system introduced by the Law of Construction Business in 1958. According to this law, every construction firm must acquire a license from the government to participate in any kind of construction business except very small construction works. There were three types of construction license, but only those firms with the general construction license are allowed to participate in large-scale construction works. The Korean government blocked entry of new construction firms by limiting the number of general licenses, under the justification of keeping unqualified firms out and improving the quality of construction business. Specifically, the Korean government stopped issuing new general construction licenses in 1975, with the result that the number of construction firms with a general license actually decreased as bankruptcy or merger and acquisition took their toll, as we can see in Figure 10.2. The government began to issue new licenses in 1989, and the number of general construction firms has increased rapidly since then.

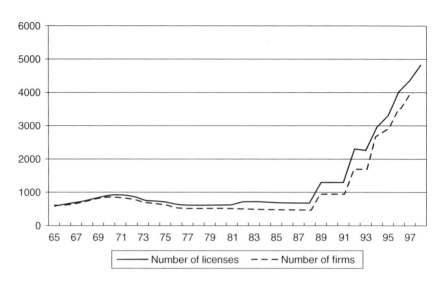

Figure 10.2 Trends of general construction licenses.

Contract ceilings

The second regulation measure is the contract ceiling system introduced in 1961, which set an upper limit to the amount of an individual construction order that a construction firm can receive. The ceiling was determined by the size of the construction firm, measured in terms of capital stock, number of engineers, and past records of construction orders received. Thus only large firms could participate in large-scale construction works, and the contract ceiling system has functioned as an entry and mobility barrier to small and medium-size firms.

Limited competition and private contracts in public construction

The third regulation was the government contract system. In principle, all public construction works should be contracted on the basis of general competition among all bidders, but, in actuality, public construction is contracted through limited competition contracts and private contracts (Sim 1994; Nam 1997). The limited competition contract limits bidding to large-scale firms, and the private contract selects a certain company and contracts with it exclusively. For example, in 1996, general competition contracts accounted for only 23.8 per cent of total public contracts, while limited competition contract and private contracts took 72.9 per cent of the total (The Construction Association of Korea, *The Statistical Yearbook of Construction Industry*, Annual Reports, 1997). Large construction firms have an advantage in limited competition and private contracts, giving the public contract system a built-in entry barrier to the construction industry. Taken together, contract licensing, ceilings on private contracts and public contract bidding procedures have created oligopolistic collusion at the top of a system of subcontracting that pushes competitive pressures down the hierarchy of subcontracting and towards the smallest units, including temporary and daily workers.

The structure and behavior of the construction market

Growth fluctuations

The development of Korean construction industry since the 1960s can be classified into three stages.

As shown in Table 10.3, the total amount of construction orders increased by 590 times, but the number of general construction firms increased by only 1.7 times during the oligopolistic growth era from 1965 to 1989. As a result, the average amount of construction orders received per firm increased by 338 times during this period.[1] The second stage is characterized by the stable growth era from 1989 to 1997, during which the government began reissuing new licenses, and the number of licenses increased. Since the amount of total construction orders increased at the same rate, the average amount of construction projects per firm was almost unchanged during this period.[2] The third stage is the era of construction crisis after the economic crisis in 1997. The average amount of construction orders received per firm dropped by 60 per cent because, while the number of firms increased by 50 per cent after the abolition of the license system in 1999, the total amount of construction works decreased by 40 per cent.

Table 10.3 Structural changes in the Korean construction industry

	1965	1989	Increasing ratio (times)
Monopolistic Growth Era (1965–89)			
Construction orders received (billion won)	27.1	16,000	575
Number of general construction firms	561	930	1.7
Average amounts per firm (million won)	48.3	17,200	338
	1989	*1997*	*ratio*
Stable Growth Era (1989–97)			
Construction orders received (billion won)	16,000	80,000	5.0
Number of general construction firms	930	3,896	4.2
Average amounts per firm (million won)	17,200	20,500	1.2
	1997	*2000*	*ratio*
Crisis Era (1997–2000)			
Construction orders received (billion won)	80,000	50,000	0.6
Number of general construction firms	3,896	6,000	1.5
Average amounts per firm (million won)	20,500	8,300	0.4

Source: Ministry of Construction and Transportation (2000a).

Table 10.4 The number of firms and value ordered by firm size, in per cent

	Number of firms				Value of orders			
	1985	*1990*	*1995*	*1998*	*1985*	*1990*	*1995*	*1998*
Total	100.0	100.0	100.0	100.0	100.0	100.0	100.0	100.0
10 persons or less	54.9	59.4	57.2	59.4	0.6	2.0	3.2	4.4
10–49 persons	27.3	24.6	27.8	27.9	3.0	8.6	12.8	15.9
50–99 persons	8.0	7.1	7.3	6.5	2.6	7.4	10.2	11.3
100–499 persons	7.3	6.9	6.7	5.6	11.5	24.1	19.2	28.9
500–999 persons	1.6	1.3	0.7	0.3	8.6	18.6	10.8	7.8
1000 persons or more	1.0	0.6	0.3	0.2	73.6	39.3	33.3	31.7

Source: Ministry of Construction and Transportation, *Yearbook of Construction & Transportation Statistics*, Annual Reports.

Concentration at the top

Overall, government regulations on the Korean construction market brought about an oligopolistic market structure and a multi-layer subcontracting system. As Table 10.4 shows, only 1 per cent of total construction firms accounted for 74 per cent of the total value of construction orders in 1985. Even though the value of construction orders in small and medium-size firms has increased under the protective policies for subcontracting firms since 1989, still the Korean construction market is characterized by its oligopolistic structure. Only 0.2 per cent of total construction firms have 1000 employees or more, but they accounted for 31.7 per cent of the total value of construction orders in 1998. Firms with 49 employees or less comprised 87.3 per cent of total construction firms but produced only 20.3 per cent of the total value of construction orders.

Multiple layers of subcontracting

Many construction firms make profits by collusion within an extensively layered subcontracting system. Large construction companies often collude among themselves to bid on government contracts and at higher prices than would be expected in normal competition. They also subcontract their construction projects received to small and medium-size firms and, in the process, get a share of those profits as their economic rents as well.

Table 10.5 shows that the ratio of the subcontracting costs of general construction firms, in the brief span of a decade, has increased from 31.5 per cent of total construction costs in 1989 to 51.8 per cent in 1998. The subcontracting system has a multi-layer form in Korea, i.e., the process of subcontracting frequently is repeated down numerous stages. In the first stage, a general construction firm itemizes the construction works and sublets these items to

Table 10.5 The ratio of subcontracting costs in total costs in general construction firms, in per cent

	1989	1992	1995	1998
Total costs	100.0	100.0	100.0	100.0
Material costs	34.3	30.1	28.5	24.5
Labor costs	21.1	17.9	13.6	11.1
Subcontracting costs	31.5	41.0	46.7	51.8
Field costs	13.1	11.1	11.2	12.5

Source: The Construction Association of Korea, *The Statistical Yearbook of Construction Industry,* Annual Reports.

medium-size special construction companies. A special construction firm further sublets its works to small subcontracting firms, and so forth, often for four or five stages. In the final stage, a small subcontracting firm contracts with *Oyaji*'s,[3] self-employed skilled workers who, in turn, employ daily workers to help them perform the itemized works (Park 1986; Hwang 1986).

Contractors who are higher in this subcontracting hierarchy, profit by squeezing the subcontractors beneath them. Superior contractors can pass over some of their production costs to the inferior subcontractors because of fierce competition among subcontractors, which forces them to accept prices that are marginal or even that engender a loss, with unfair terms of payments (Lee 1994; Bak 1997; Presidential Commission on Anti-Corruption 2000).

The superior contractor also avoids various responsibilities as an employer by passing them along to the subcontracting firms. The burden shifts continuously to subcontractors further down the hierarchy and finally to the daily workers, resulting in low wages, unstable employment, and poor working conditions.

In order to prevent excessive subcontracting, the Korean government has enacted a series of regulations on the subcontracting system. According to the Fair Trade Act on Subcontract, enacted in 1984, construction firms cannot sublet the major part of construction works to the subcontractors. The subcontractors also cannot resublet the construction works to other subcontractors without the permission of the original issuers. The original contractors are enjoined to treat subcontractors in a fair manner, i.e., to provide fair pricing and minimize payment delays.

However, these regulations are seldom respected among construction firms. There are many cases of illegal practices, such as subcontracting whole construction works, resubcontracting, lending licenses to an unlicensed subcontractor, fictitious employment of subcontractors to avoid the regulations, asking extremely low prices from subcontractors, and delay of payment for several months. The government has not actively monitored and prosecuted the violation of the law, so these illegal practices are still widespread among construction firms (Bak 1997; Presidential Commission on Anti-Corruption 2000).

The performance of the construction industry

An oligopolistic market structure and a multi-layer subcontracting system have negative consequences on the performance of the Korean construction industry. Illegal construction practices and the low quality of construction products represent the most important negative effects. The low pricing for subcontracted work leads to shoddy construction works. Also the intense competition for subcontracts brings about corruption; many cases of bribery have been reported.

Although the Korean construction industry has experienced high growth, its competitiveness is still low. According to the Korea Institute for Industrial Economics (2000), the technological level of the Korean construction industry was only two-thirds of that of the advanced countries in 1998. Competitiveness is particularly low in the areas of planning and management. For example, while the Korean construction market ranks ninth in the world by size, engineering ability in the Korean construction industry ranks only twenty-fifth (Ministry of Construction and Transportation 2000a).

Figure 10.3 compares the productivity in construction with that in manufacturing, as measured by the value-added per employee in 1995 constant prices. The productivity of construction industry was twice as high as that of manufacturing in the 1970s. This productivity gap was maintained through the 1980s but narrowed until the middle of the 1990s. Since then, while manufacturing productivity has increased rapidly, construction productivity has stagnated, and the productivity gap was reversed in the second half of the 1990s. By 1999, the productivity of manufacturing was 40 per cent higher than that of construction.

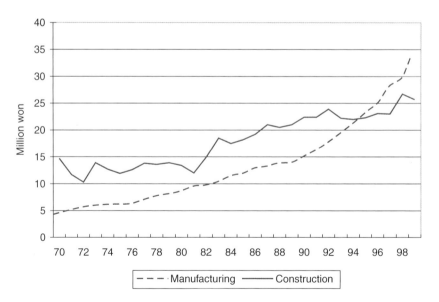

Figure 10.3 Trends of productivity in construction and manufacturing, in 1995 constant prices.

Note: Value-added per employee.

FREE-FALL IN THE CONSTRUCTION LABOR MARKET

Regulations on the labor market

Weak regulations on labor standards

Regulations of labor standards are very weak in the Korean construction industry because of incomplete labor laws and insufficient oversight. All industries in Korea, including the construction industry, are subject to the Labor Standards Act, enacted in 1953, which regulates wages, work hours, and working conditions. However, the law is not quite appropriate for the construction industry because it was enacted focusing on the regular workers in manufacturing firms.

Korean labor law permits a short-term employment contract of less than 1 year without any restrictions. Recently, the numbers of temporary and daily workers in construction have increased as business firms substitute them for regular workers to gain more employment flexibility and to reduce labor costs. Temporary workers, defined as those whose employment period is more than one month but less than one year, accounted for 20.6 per cent of total employees for the whole industry in 1999. Daily workers, whose employment period is less than one month, accounted for 11.3 per cent of total employees in the same year. Therefore, what is atypical labor in other countries accounts for 31.9 per cent of total workforce and 51.7 per cent of total wage earners in 1999. This ratio of atypical labor is very high compared to the ratio in other OECD countries (OECD 1999). Temporary and daily workers in Korea suffer job insecurity, low wages, and poor working conditions, and usually they are not protected by social insurance or trade unions (Korean Confederation of Trade Unions 2000a).

Overall, Korean workers also have the longest work hours among OECD countries. OECD (2000) reported that the average annual work hours of a Korean worker amounted to 2497 hours in 1999 – about 600–900 hours more than that of most European countries. Korean labor law permits 44 work hours per week as the standard, and this standard may be extended by 12 hours more with worker consent. However, construction firms rarely keep this regulation strictly, and long work hours prevail in the Korean construction industry. A recent survey from a construction labor union shows that the average work week of field workers is 68.7 hours, or 24 hours longer than legal standard. According to this survey, the average work hours on a weekday amounted to 11 hours and 20 minutes – from 7:41 a.m. to 7:04 p.m. – in 1999. The work hours on Saturday amounted to 8 hours and 20 minutes – from 7:41 a.m. to 4:04 p.m. (Korean Federation of Construction Industry Trade Unions 2000). Moreover, construction workers cannot enjoy a regular holiday. The Labor Standards Act provides one paid holiday per month in addition to Sundays if there is no absent day in that month. This law also provides 10–20 annual paid holidays, in proportion to a worker's tenure in a company. However, the work days of daily workers in construction are too intermittent to get a monthly holiday, and the length of service for a company is too short to entitle them to the paid annual holidays.

Regarding the wages of construction workers, the Minimum Wage Act of 1986 provides the only regulation. However, because the minimum wage is so low – only 24.2 per cent of the average wage in 1999 – only a small number of the total workers – 0.4 per cent – effectively benefited from the minimum wage law in that year (Korea Labor Institute 1999).

Exclusion from social security systems

Although there are four major social insurance systems in Korea – i.e., medical insurance, pension insurance, unemployment insurance, and industrial accidents insurance – most daily workers in the construction industry are covered only by industrial accidents insurance. As can be seen in Table 10.6, most construction workers are not qualified for workplace medical insurance because of the requirement of two months continuous employment in the same workplace. Daily workers can join a community medical insurance program, but its premium is more expensive because there are no contributions from employers to the program.

A worker has to serve more than 3 months continuously in the same company to quality for pension insurance and one month to qualify for unemployment insurance. As a result, most daily workers are not covered by these forms of insurance either.

In July 2000, the Industrial Accidents Compensation Insurance Act, which had applied only to business firms with five or more employees, expanded to cover all business firms. However, it does not cover small construction projects.[4] Recently, the Korean government expanded the coverage of social insurances, but most daily workers are still excluded from them. For example, a special law

Table 10.6 The eligibility of social insurances

		Medical insurance	Pension insurance	Unemployment insurance	Industrial accidents insurance
Regular worker	5 employees or more	○	○	○	○
	4 employees or less	X	X	○	○
Atypical worker	Daily worker	□ (over 2 mo.)	□ (over 3 mo.)	□ (over 1 mo.)	○
	Temporary worker	□ (over 2 mo.)	□ (over 3 mo.)	□ (over 1 mo.)	○
	Part-time worker	X	X	□ (over 80 hours per mo.)	○

Source: KCTU 2000b.

Note
○ = fully eligible, □ = partly eligible, X = non-eligible.

aims to give a retirement allowance to construction workers. Under the Construction Employment Improvement Act of 1996, the Retirement Credit System established a mutual aid system for building retirement credits. Now workers may manage their own records with a welfare card to which employers attach stamps in proportion to the number of days worked. Upon the worker's retirement, a benefit association pays a retirement allowance according to the work records. However, the system is based on the voluntary participation of employers, except in the case of very large construction works, and its effect has been limited.[5] Thus, there is little in the way of wage and hours regulations or unemployment insurance and pensions supporting labor standards in Korean construction.

Lack of protection through collective bargaining

Usually construction field workers in Korea are neither protected by a labor union nor covered by a collective bargaining agreement. The total number of trade union members in the construction industry is approximately 20,000, just one per cent of the total construction workforce. In addition, in the almost complete absence of collective agreements on an industry-wide or even regional basis, construction trade unions have little power to check unfair dismissals or arbitrary labor management by employers.

The Korean government had strictly suppressed independent labor movements until 1987 under the justification of rapid economic development. Most trade unions were organized not by industry or trade, but rather by enterprise. The government authorized the Federation of Korean Trade Unions (FKTU) as the only legal national center. After the general strike of 1987, many democratic unions were newly organized. They formed the Korean Confederation of Trade Unions (KCTU) as their national center in 1995.

However, organizing construction workers was difficult because construction projects are intermittent and dispersed throughout the nation, and because most construction workers are employed temporarily. The enterprise unions of white-collar employees and engineers in the head office of construction firms organized the Korean Federation of Construction Trade Unions (KFCTU) as their umbrella organization in December 1989. Blue-collar workers in the construction industry have organized regional trade unions since 1987. These regional unions formed the Korean Construction Workers Union (KCWU) in April 1989, but this coalition did not acquire legal status until 1998.

Since 1997, with the construction industry in crisis, many construction unions have lost members because of mass unemployment. To overcome this labor union crisis, white-collar and blue-collar workers in the construction industry joined together in a unified labor organization, the Korean Federation of Construction Industry Trade Unions (KFCITU), formed by the KFCTU and the KCWU in December 1999. The KFCITU is affiliated with the KCTU. White-collar workers in the public sector also have confronted mass layoffs since 1997, so they organized the Korean Public Sector Construction Trade Unions

(KPCTU) in 1998. This organization is affiliated with the FKTU. As we can see in Table 10.7, the total number of union members in three construction unions was only a little over 21,000 persons in July 2000.

Collective bargaining is not very developed in the Korean construction industry (Members of KFCITU 2000). In the case of enterprise unions, each union under the KFCITU and KPCTU bargains directly with the individual company. The individual enterprise unions may delegate collective bargaining directly to an industrial federation of unions under the new labor law. Three of the enterprise unions delegated collective bargaining to the KFCITU in 2000. Under the delegation bargaining system, representatives from the KFCITU and the individual trade union sit together and bargain with representatives from the company. The KFCITU has tried to organize a German-style industry-wide trade union and conduct unified bargaining with the employers association, but this effort faces strong opposition from employers.

For the regional trade unions, the KCWU has not yet succeeded in collective bargaining with employers, except a few cases. Employers commonly refuse to bargain with the regional unions, labeling them "outsiders." Thus, most blue-collar construction workers are employed through individual employment contracts without the involvement of regional trade unions. The regional trade unions request a standard form of individual employment contract to the

Table 10.7 The number of construction trade unions and union members

	KCTU		FKTU
	KFCTU	*KCWU*	*KPCTU*
1987	8 (3,607)	n.a.	
1988	21 (9,339)	n.a.	
1989	29 (16,591)	13 (6,000)	
1990	29 (16,591)	n.a.	
1991	30 (17,341)	n.a.	
1992	31 (17,461)	5 (1,000)	
1993	34 (18,024)	n.a.	
1994	38 (18,567)	n.a.	
1995	43 (17,264)	5 (1,000)	
1996	53 (18,863)	6 (1,000)	
1997	50 (18,945)	7 (1,050)	
1998	48 (15,130)	22 (1,500)	3 (5,400)
1999	45 (11,202)	32 (1,692)	8 (5,429)
2000.7*	49 (12,000)	39 (3,000)	8 (6,509)

Sources: KCTU; FKTU; KFCITU; KPCTU.

Notes

n.a. = not available; ( ) notes the number of union members.

* The KFCTU and the KCWU merged into the KFCITU in 1999, but the two organizations remain as semi-independent organizations within the KFCITU.

employer. Sometimes they withdraw union members from construction sites where employers reject collective bargaining or violate a collective agreement. However, their intervention is typically quite limited due to the low percentage of workers in the regional unions. Thus, to this point in the development of Korean construction, collective bargaining has not substituted for the absence of state regulations of labor standards and social insurance.

The structure of the construction labor market

The multi-layered subcontracting system under the oligopolistic construction market determines the structure of the construction labor market. Korean construction firms usually do not employ field workers directly. Instead, they employ workers indirectly in the form of daily workers under the subcontracting system because it increases employment flexibility and offloads the responsibility of labor control to subcontractors.

An *Oyaji* attempts to maximize his profits by lowering wages and other expenses as far as he can. He performs many functions such as mobilization and disposition of daily workers, process management, wage control, and job training. An *Oyaji* and daily workers make a small team together on the basis of their informal job network. The small team is very flexible; daily workers instantly gather and scatter depending on the job situation. Usually an *Oyaji* has connections with more than 20 daily workers and can mobilize them when and where they are needed. His network consists of relatives, members of his neighborhood, and acquaintances of long standing (Korea Labor Institute 1998). Their employment contract – usually an oral agreement only – terminates automatically when a construction project is finished. There is no job stability or job promotion for daily workers. A daily worker can hope to become an *Oyaji* someday, after acquiring skills and accumulating decent capital, but it is very difficult to realize this dream because of insufficient capital accumulation and connections with construction firms.

While most daily workers had come from rural areas until the 1970s, they began to be mobilized from urban areas after the 1980s. Besides the *Oyaji*-based system, many construction workers are now mobilized through daily labor markets or temporary labor agencies. In the case of daily labor markets, small construction firms or *Oyaji*'s make oral contracts on the spot with the daily workers based on a daily payment system. There are around 60 daily labor markets across the nation.

In 1998, the Korean government introduced the Worker Dispatch Act, which legalized temporary labor agencies. Although this law excluded the construction industry from this legalization, out of concern that the agencies would exploit construction workers, many agencies illegally dispatch workers to the construction companies anyway, and the government does not actively prosecute these actions. Most job seekers through temporary labor agencies are unskilled construction workers.

The performance of the construction labor market

The low quality of construction labor

The oligopolistic market structure and the multi-layer subcontracting system in the construction industry have negative effects on the performance of the construction labor market. First, the quality of construction workers has degenerated since the 1980s. Qualified workers are not willing to enter the construction industry because of unstable employment, low wages, and poor working conditions.

As we can see in Table 10.8, the average age of construction workers is getting older. In 1983, young workers in their teens and twenties accounted for 26.2 per cent of total construction workers. By 1999, however, the proportion of young workers had dropped to 18.4 per cent. At the other end of the age range, workers of 50 or more years accounted for only 14.5 per cent of total construction workers in 1983 but had increased their percentage of the construction workforce to 19.0 per cent by 1999. When we compare the age composition in the construction industry with that of total employment, it is evident that the average age of construction workers is older than that of workers as a whole. Workers aged 24 or younger decreased to only 5.9 per cent of total construction workers in 1999, whereas their proportion of total non-agricultural employment was 9.9 per cent. Because of job insecurity and hard working conditions, young workers have begun to avoid entering the construction industry.

Table 10.8 also shows educational attainment among construction workers. With the general upward trend in educational level, the average level of education of construction workers has increased dramatically since the 1980s. If we compare the distribution of educational attainment among construction workers

Table 10.8 The composition of construction workers by age and education, in per cent

	1983	1986	1989	1992	1995	1999	1999* (Total)
Age							
24 years or less	9.8	6.5	6.2	8.6	9.1	5.9	9.9
25–29 years	16.4	15.2	11.8	12.0	12.7	12.5	13.8
30–39 years	30.5	34.0	32.7	34.1	33.2	33.7	31.4
40–49 years	28.7	28.9	28.2	25.7	25.3	29.1	25.9
50–59 years	12.1	13.6	17.9	16.5	15.1	14.3	13.3
60 years or over	2.4	1.7	3.2	3.3	4.5	4.7	5.6
Education							
Middle school or less	64.9	60.6	60.5	47.8	40.1	31.1	27.0
High school	24.7	28.4	30.8	38.1	44.7	48.4	46.6
College or over	10.3	11.1	8.6	14.1	15.1	20.4	26.5

Source: Korea National Statistical Office, *Annual Report on the Economically Active Population Survey*, Annual Reports.

Note
* Total non-agricultural employment in 1999.

with that of total employment, however, it is clear that the construction industry is staffed with less-educated workers.

The low skill level of construction workers

The skill formation process of construction workers is very weak and informal in Korea. Construction daily workers acquire trade skills usually from senior workers on the job without a chance to receive any systematic job training. They start construction works as assistant laborers and become full craft workers after 3–5 years. However, there is no formal qualification for promotion to a craftsperson.

Table 10.9 shows that most construction workers have not received any kind of formal job training. The ratio of construction workers with any experience of formal job training is only 12.2 per cent of total construction workers in 1997. Among those who have ever received formal job training, 40.3 per cent received training from the vocational training system in companies, 34.3 per cent from private institutes, and 25.4 per cent from public vocational training institutes. Thus, it is evident that the government plays only a minimal role in the vocational training of construction workers.

There are several reasons why vocational training for construction workers is very weak in Korea. First, construction firms are not interested in fostering skilled workers because the mobility of construction workers is so high. Second, construction workers have almost no incentive to participate in the programs. Most of them are temporary workers, so better training is unlikely to lead to promotion or higher wages. Third, the government has shown little interest in fostering skilled workers in construction in the past. Fourth, construction trade unions are too weak to be involved in the training programs (Kim and Sim 1999; Sim 2000).

The main official source of construction skills training in Korea is the vocational education and training programs that are financed by the employment insurance system. These programs support the employers who provide job training to their workers and consist of entry training for new workers and advanced training for existing workers. However, these programs are rarely applied to the construction industry because most construction workers are daily workers and thus are not covered by the employment insurance system.

Table 10.9 Vocational training experience among construction workers, in per cent

	1986	1992	1997
Yes	7.2	13.2	12.2
Private Institute	(50.0)	(47.7)	(34.3)
Public Vocational Training	(25.0)	(26.7)	(25.4)
In-house Vocational Training	(25.0)	(23.6)	(40.3)
No	92.8	86.8	87.8

Source: Korea National Statistical Office, *Report on the Employment Structure Survey*, various years.

Moreover, there are few applicants for the entry training because young workers are not attracted to work in the construction industry (Sim 2000). Consequently, most daily workers learn skills informally from senior workers or from *Oyaji's*. This informal process of skill formation presents many problems, such as the restriction of learning opportunities, the difficulty of training new techniques, and the narrow range of skill delivery. Therefore, it cannot contribute much to the improvement of skills of construction workers and the quality of final products (Sim 2000).

Low job security in the construction labor market

Job security is very low in the construction labor market. Most construction workers are temporary or daily workers with a short employment contract period. Table 10.10 shows that the ratio of daily workers whose period of employment is less than one month increased from 52.2 per cent of total construction workers in 1983 to 59.9 per cent in 1989. Although the ratio had decreased since 1989, it increased again after the economic crisis of 1997 to 42.9 per cent in 1999.

Also, the ratio of temporary workers whose job tenure is less than one year increased from 9.5 per cent in 1983 to 14.8 per cent in 1999. Therefore, it is estimated that temporary and daily workers amounted to 57.7 per cent of total construction employment in 1999.

Korean labor law permits employment contracts of less than 1 year for whatever reason and in whatever form. The labor unions call for new regulation on short-term employment contracts, but employers strongly oppose any changes. Up to now, the government has taken no action to stem the proliferation of temporary workers.

Table 10.10 The ratio of workers by type of employment in construction industry, in per cent

	1983	1986	1989	1992	1997	1998	1999
Employer	2.9	3.7	3.9	6.0	9.6	21.9*	21.2*
Self-employed	6.5	7.7	5.4	7.4	8.5		
Family workers	0.3	0.7	0.8	1.4	0.9		
Regular workers	28.4	33.9	30.0	38.5	42.7	22.8	21.1
Temporary workers	9.5					14.9	14.8
Daily workers	52.2	54.0	59.9	46.7	38.4	40.4	42.9

Source: Korea National Statistical Office, *Annual Report on the Economically Active Population Survey*, various years.

Notes
* includes self-employed and family workers.
Regular workers = the period of employment contract is not specified or more than one year.
Temporary workers = the period of employment contract is less than 1 year but longer than 1 month.
Daily workers = the period of employment contract is less than 1 month.

Low wages in the construction industry

The wage of construction workers has decreased compared to other industries since the 1980s. In the past, construction work was characterized as a hard-working but well-paid job, and wages in construction industry were higher than those in most other industries. The wage differential was a kind of premium for the unstable employment and poor working conditions in the industry.

As is seen in Figure 10.4, monthly pay for construction workers was 75.7 per cent higher than that of manufacturing workers in 1980. Since the middle of the 1980s, however, the wage differential has narrowed to the point where monthly pay for construction workers in 1999 was only 14.6 per cent higher than that of manufacturing workers. The reduction of the wage differential is a reflection of the low productivity, low education, low skill, and low bargaining power in the construction labor market.

Furthermore, the average level of construction wage conceals internal inequality in the construction labor market. The wage level of field workers in the construction industry is much lower than that of white-collar or technical construction workers. Table 10.11 shows that, standardizing the wage level of engineers at 100, the indices of that of craftspersons and laborers are only 75.2 and 69.4, respectively, in 1999. While the wage differential between engineers and craftspersons has been reduced since 1990, it has been increased between engineers and laborers since 1990. The wage differential between engineers and temporary craftspersons has not changed much since 1990.

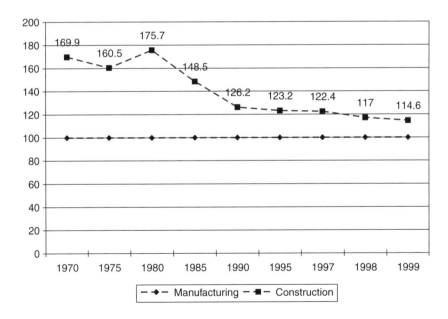

Figure 10.4 Relative wages of construction and manufacturing.

Table 10.11 Wage differentials by occupation in the construction industry

	Administration and others	Operatives				
		Total	Engineers	Crafts-persons	Temporary crafts-persons	Laborers
1990	101.6	82.1	100.0	54.6	85.6	81.4
1995	90.7	66.7	100.0	71.1	86.2	53.1
1997	82.5	83.2	100.0	75.5	97.9	70.0
1998	92.0	79.6	100.0	75.2	89.8	69.4

Source: Korea National Statistical Office, *Report on the Construction Work Survey*, various years.

Thus, regulation of the Korean construction labor market provides limited insurance against accident, no insurance against unemployment or old age, almost no collective bargaining, a relatively ineffective state-mandated minimum wage, and almost no limit on hours worked. Unions currently account for approximately 1 per cent of the construction labor force and are consequently not up to the task of replacing state regulations with collective agreements. Neither contractors nor the State nor collective bargaining has succeeded in establishing a system of formal training. Skill formation is correspondingly low, and labor productivity and wages consequently suffer. These factors are a recipe for job insecurity and a falling assessment of construction work as a viable alternative for young people.

RECENT DEREGULATION POLICIES AND THEIR IMPACTS ON THE CONSTRUCTION INDUSTRY

The Korean government began to deregulate the construction industry under the overall deregulation policies of the second half of the 1990s, accelerating after the economic crisis in 1997. The International Monetary Fund (IMF) asked the Korean government to implement a series of neo-liberal reforms as a condition of an emergency rescue loan. Following this request, the Korean government has implemented a series of reform packages including deregulation and opening to foreign investment.

As we saw earlier, the biggest change in the construction industry has been the abolishment of the contractor license system and its replacement with a non-restrictive registration system in 1999. It is estimated that the number of general construction firms increased to 6000 by the end of 2000 because of the abolition of the license system.

The contract ceiling system has been changed also. After abolishing the contract ceiling system in 1997, the Korean government introduced the public announcement system. As a result, an ordering party can now freely choose any construction firm without considering the ceiling of the firm. Instead of a ceiling, the limit for a single construction order is determined on the basis of publicly announced competence. The competence of a construction firm is evaluated by referring to the construction records, capital stock, the level of technology, reputations, etc. (Nam 1997). However, this announced competence is for reference purposes only and not a form of regulation.

Furthermore, the government abolished the limited competition system and introduced a lowest-bidder-wins system into public contracts. Under the new system, any company bidding the lowest price can get an order from the public sector (Ministry of Construction and Transportation 2000b). This institutional change aims to reduce the implicit collusion practices among the large construction firms.

In 1994, the Korean government began to allow foreigners to invest in the Korean construction market and has expanded areas in which they can participate since then. Recently foreign construction firms have participated in large-scale construction works, such as the Inchoen International Airport and the Seoul–Busan Express Railway in various areas such as design, engineering, superintendence, and safety evaluation (The Construction Association of Korea 1994; Kim and Lee 1994).

The Korean government introduced a series of deregulation measures in the construction labor market in accordance with policies designed to increase the flexibility of labor. First, in 1998, the government introduced the Worker Dispatch Act and legalized redundancy dismissals. Since then redundancy dismissals of white-collar workers have rapidly increased in the construction labor market. As we have seen earlier, worker dispatch by temporary labor companies is still not allowed in the construction industry, but actually many construction firms now use these illegal companies to recruit daily workers.

The Korean government also introduced a series of deregulation measures in the field of industrial safety and health in the latter half of the 1990s. First, the government offices and staff in charge of industrial safety and health affairs have decreased in accordance with the overall trend toward smaller government. These changes make it more difficult to monitor and prosecute violations of industrial safety law by employers. Second, the obligation of employers to employ a certain number of licensed experts in the field of industrial safety has been deregulated. Both the number of persons and the level of their qualification have been lowered. For example, large companies are no longer required to employ medical doctors. Third, the obligation of employers to educate and train their workers on industrial safety has been abolished.

These deregulation measures have influenced both construction product markets and the working conditions of construction workers. As entry barriers have been lowered, the number of construction firms has increased rapidly so that the construction market turned highly competitive. However, measures that would have ensured fair competition have not been sufficiently developed. As the result, there is now a free-for-all, cut-throat competition in the construction industry. Many construction firms bankrupted because of fierce competition, and the spillover from the economic crisis of 1997 has only aggravated the situation.

The price-cutting competition among construction firms has direct impacts on the wages and working conditions of construction workers. More and more companies rely upon subcontracting and temporary labor agencies to cut costs and increase their employment flexibility. They spend less money on the training of their workers or safety investments.

These changes disturb the existing balance in the construction labor market and accelerate job insecurity, low wages, and worsening working conditions of construction workers. The fierce competition also has a long-term impact on the productivity of the industry. Because of lower investments in human capital and the deteriorating situation in the construction labor market, young workers avoid entering the construction industry, and existing skilled workers leave the industry to seek more promising jobs. This trend will have long-term impacts on the construction industry.

The Korean government recently acknowledged these problems and introduced some protective measures for construction workers. As the number of atypical workers has increased rapidly, the government is considering some protection measures for them, including regulation of the employment contract term. Also, the coverage of social insurance has been expanded to include more construction workers. For example, a retirement credit system for construction workers was introduced in 1996. Four major social insurance systems also have expanded to cover more construction workers. Recently, construction trade unions also have tried to integrate across individual unions and to enforce collective agreements by organizing an industry-wide trade union. However, these protective measures are still very limited in scope and depth and their effects have not materialized yet.

CONCLUSION

Korean construction presents a case where the product market has historically been closely regulated through contractor licensing and other regulations, while the labor market has not been regulated either through state regulations or through collectively bargained contracts. Unions account for only 1 per cent of the construction labor force. The result has been a firm structure characterized by oligopolistic collusion at the top and a multi-layered subcontracting system that pushes competitive pressures downward onto the shoulders of temporary and unprotected workers. The historical result has been a labor force with little training and low productivity. Recently, the product market has been deregulated. The economic crisis of 1997 has combined with product market deregulation to redouble the pressure on temporary and daily workers who enter construction with no promotional prospects, no job security, no social security benefits except industrial accident insurance, no formal training prospects, and no union protection.

The Korean government has not properly regulated long work hours, violations of the Labor Standards Act, or industrial accidents in the construction industry. Recently, the government expanded the coverage of unemployment insurance and introduced a Retirement Credit System for construction workers. However, the effects of these measures are not clear because of the conditional clauses in those measures. Recent deregulation policies in the Korean construction industry have made things worse. As the entry barriers have been lowered, the number of construction firms has increased rapidly, and the construction industry has become highly competitive. Price-cutting competition among construction firms has direct impacts on the wages and working conditions of construction workers. Increasingly, companies have relied upon subcontracting and temporary labor agencies to cut costs and gain employment flexibility. They spend less money on the training of their workers or investments in safety measures.

These changes disturb the existing balance in construction labor market, and accelerate job insecurity, low wages, and worsening working conditions of construction workers. The fierce competition also has a long-term impact on the productiveness of the construction industry. Because of lower investments in human capital and the deteriorating situation in the construction labor market, young workers avoid entering the construction industry, and existing skilled workers leave the industry for more promising jobs. This trend will have long-term impacts on the construction industry.

To solve these problems, the Korean government needs to carry out more actively a series of legal and institutional reforms to improve job insecurity, poor working conditions, and low wages of construction workers. Instead of one-sided deregulation on the construction industry, the Korean government could and should introduce a series of re-regulation measures that would ensure fair competition. Also the government should reinforce the protection of irregular workers and strictly apply the Labor Standards Act to construction workers. Finally, an industry-wide collective bargaining agreement is also very important to secure the fairness of the construction labor market in the future. The construction unions should strengthen their efforts to acquire such an industry-wide collective agreement.

NOTES

1 During this period, the general price level increased by 10 times, so the increase rate in real terms is 34 times.
2 During this period, the general price level increased by only 11.7 per cent.
3 In the Japanese language, an *Oyaji* is a "father" or a "head of work team." The Korean construction industry began during the Japanese colonial period (1910–45), and many Japanese terms are still in use today in construction industry.
4 The law is not applied to the construction works of less than 20 million won or 330 square meters.
5 Construction firms that receive public construction orders of more than 10 billion won or that build more than 500 units of housing are required to participate in this system.

REFERENCES

Bak, J. K. (1997) *A Study on the Unfair Subcontracting System in the Korean Construction Industry*, Seoul: Hanyang University Press.

The Construction Association of Korea (1994) *The Cost Analysis of Construction Works*, mimeographed, The Construction Association of Korea.

The Construction Association of Korea (various years) *The Statistical Yearbook of Construction Industry, Annual Reports*, Seoul: The Construction Association of Korea.

Hwang, I. J. (1986) "A study of the employment structure in the Korean construction industry," *A Study of Korean Society*, no. 4, Seoul: Hangil Press.

Kim, H. and Sim, K. B. (1999) *A Study of the Employment Structure of Construction Workers*, Seoul: Korea Labor Institute.

Kim, H. S. and Lee, S. J. (1994) *A Study of the Globalization of the Korean Construction Industry*, Seoul: Korea Research Institute for Human Settlements.

Kim, Y. J. (1997) *The Long-term Strategy of Supply for Construction Materials and Manpower*, Seoul: Korea Research Institute for Human Settlements.

Korea Institute for Industrial Economics (2000) *The Korean Industry*, Seoul: Korea Institute for Industrial Economics.

Korea Labor Institute (1998) *The Life and Employment Structure of Construction Daily Workers*, Seoul: Korea Labor Institute.

Korea Labor Institute (1999) *Korea Labor Institute Labor Statistics*, Seoul: Korea Labor Institute.

Korea National Statistical Office (various years) *Annual Report on the Economically Active Population Survey*, Seoul: Korea National Statistical Office.

Korea National Statistical Office (various years) *Yearbook of Construction Statistics, Annual Report*, Seoul: Korea National Statistical Office.

Korea National Statistical Office (various years) *Report on the Construction Work Survey, Annual Report*, Seoul: Korea National Statistical Office.

Korea National Statistical Office (1986, 1992, 1997) *Report on the Employment Structure Survey*, Seoul: Korea National Statistical Office.

Korean Confederation of Trade Unions (KCTU) (2000a) *A Study of the Atypical Workers Organization*, Seoul: KCTU.

Korean Confederation of Trade Unions (KCTU) (2000b) *The Social Welfare Policy*, Seoul: KCTU.

Korean Federation of Construction Industry Trade Unions (KFCITU) (1999) *Conditions in the Vocational Education Institute*, Seoul: KFCITU.

Korean Federation of Construction Industry Trade Unions (KFCITU) (2000) *A Study of Working Hours in the Construction Industry and Policy Issues*, Seoul: KFCITU.

Lee, H. H. (1994) *A Study of the Improvement of the Subcontracting System in the Korean Professional Construction Business*, Seoul: Sungsil University Press.

Members of Korean Federation of Construction Industry Trade Unions (2000) Interview by author, 29 August, Seoul.

Ministry of Construction and Transportation (various years) *Yearbook of Construction & Transportation Statistics, Annual Report*, Seoul: Ministry of Construction and Transportation.

Ministry of Construction and Transportation (2000a) *The Revitalization and Structural Reforms in Construction Industry*, Seoul: Ministry of Construction and Transportation.

Ministry of Construction and Transportation (2000b) *The Structural Reforms in the Korean Construction Industry in the Era of the New Millennium*, Seoul: Ministry of Construction and Transportation.

Nam, J. K. (1997) *Handbook of Construction Industry Law*, Seoul: Gigong Press.

Organisation for Economic Co-operation and Development (OECD) (1999) *Employment Outlook*, Annual Report, Paris: OECD.

Organisation for Economic Co-operation and Development (OECD) (2000) *Employment Outlook*, Annual Report, Paris: OECD.

Park, S. H. (1986) "The characteristics of the labor market in large construction firms," *A Study of Korean Society*, no. 4, Seoul: Hangil Press.

Presidential Commission on Anti-Corruption (2000) *A Study on the Prevention of Corruption in Subcontracting System*, Seoul: Presidential Commission on Anti-Corruption.

Sim, K. B. (2000) *The Informality of the Korean Construction Labor Market and the Limit of Skill Formation*, Seoul: Korea University.

Sim, Y. S. (1994) *The Construction Industry*, Seoul: Woongjin Press.

Index

turbulence - construction is seasonal, + susceptible to postponable demand.

this instability contributes to the high labour turnover in the industry + the proliferation of subcontracting.

Every locale has a construction industry. Construction also more services centred than product centred since the customer needs the product in a specific location. ie it cannot be built elsewhere + moved. notwithstanding some pre-fab + modularization

all of this leads to a general chaos in construction that ultimately, in the case of skilled labour, compel a country to take the low road or the high road.

* leading to wide ranges + variation in the size of construction firms. this adds to the instability in the labour force because they must move from project to project